BEYOND POSTPROCESS AND POSTMODERNISM

ESSAYS ON THE SPACIOUSNESS OF RHETORIC

BEYOND POSTPROCESS AND POSTMODERNISM

ESSAYS ON THE SPACIOUSNESS OF RHETORIC

Edited by

Theresa Enos
University of Arizona

Keith D. Miller
Arizona State University

Jill McCracken, Assistant Editor
University of Arizona

LAWRENCE ERLBAUM ASSOCIATES, PUBLISHERS
2003   Mahwah, New Jersey   London

Camera ready copy for this book was provided by the editors.

Lawrence Erlbaum Associates, Inc., Publishers
10 Industrial Avenue
Mahwah, NJ 07430

Cover design by Kathryn Houghtaling Lacey

Library of Congress Cataloging-in-Publication Data

Essays on the spaciousness of rhetoric / [edited by] Theresa Enos,
    Keith D. Miller.
        p. cm.
    Includes bibliographical references and index.
ISBN 0-8058-4407-4 (cloth : alk. paper)
ISBN 0-8058-4408-2 (pbk. : alk. paper)
I. English language—Rhetoric—Study and teaching. 2. Report writing—
    Study and teaching (Higher). I. Enos, Theresa. II. Miller, Keith D.
PE1404 .E8 2002
808'.042'071—dc21   2002024272
                                                            CIP

Books published by Lawrence Erlbaum Associates are printed on acid-
free paper, and their bindings are chosen for strength and durability.

Printed in the United States of America
10 9 8 7 6 5 4 3 2 1

Contents

ঌ৵ ঌ৵

Part III: Parallels, Extensions, and Applications

Introduction

☙Some time after the newly fascinating discipline of rhetoric and composition studies began to emerge in the 1970s and early 1980s, it seemed to undergo dizzying changes. Enshrined in a mantra—"Teach process, not product"—process pedagogy initially appeared quite revolutionary and far superior to the current-traditional paradigm. But while enthusiasm for expressivism remained high in some quarters, Andrea Lunsford, Lisa Ede, and others began insisting that writing is collaborative. Advanced by John Hayes and Linda Flower, cognitive approaches to understanding the composing process gained popularity before interests turned elsewhere. Sharon Crowley and Victor Vitanza began pondering and interpreting the works of Michel Foucault, Jacques Derrida, and other European and American postmodernists. Cheryl Glenn, C. Jan Swearingen, and other feminists interrogated and challenged patriarchal dimensions of classical rhetoric, the current-traditional paradigm, and process pedagogy alike. James Berlin explained the importance of economic strata in defining Americans' minds and lives. Studies of the intertwining strands of race, class, and gender further enriched and complicated the discipline. Certainly, rhetoric and composition studies is now—and perhaps has always been—a complex field characterized by agreements and tensions, as bodies of thought crash, merge, and shift like the tectonic plates of the earth's surface.

The editors and contributors to this volume seek to illuminate and complicate many of the tensions in the field and thereby to contribute to postprocess pedagogy and post-postmodernist rhetoric. If we succeed, we will also honor the late Jim Corder, whose body of work reconciles opposites, provides a sustained search for ethos, offers a prophet's call for the commodiousness of language and voice, and attempts to answer the ubiquitous question of why people listen to some but not others.

Although this collection of original essays is not meant to be in the genre of Festschriften, it is offered in the spirit of Corder and his work. The issues he wrestled with are just as relevant today—perhaps more so after the innumerable issues arising out of the terrorist attacks on the United States on September 11, 2001. Throughout the process of brainstorming, proposing, writing, editing, and compiling this book, the editors and contributors have spotlighted issues that they think are as relevant now as they were in the past and that they hope will be generative for the future.

Like Corder, contributors venture into uncatalogued places. Focusing on what they deem his most useful ideas, they explore his uncategorizable mix of West Texas expressivism, rhetorical theory, and process and postprocess composition theory. Among other topics, they examine his adaptation of French phenomenology and existentialist

vii

psychology; his subversion of the binary of argumentation and expressivism; his highly expansive redefinitions of conceptions in Aristotelian and Ciceronian rhetoric; the possibility that his rhetoric is radical and feminist; his undermining of all taxonomic divisions of discourse; his intense wrestling with European postmodernism; his commitment to comity and call for commodious language; and his explication and practice of gentle, dialogic persuasion.

In Part I, Historical Context—Rhetoric and Composition Studies, Janice Lauer in "The Spaciousness of Rhetoric" contextualizes and highlights the emergence of a fledgling discipline during the 1970s. She analyzes major concerns that arose at rhetoric and composition conferences—several of which she helped spearhead—and notes key publications.

Then in Part II, Theory-Building and Critiquing Corderian Rhetoric, James S. Baumlin, who coauthored four essays with Corder, in "Toward a Corderian Theory of Rhetoric," identifies a "Corderian rhetoric" that evolves from a heroic phase to a tragic phase. Baumlin maintains that Corder bases his rhetoric on existentialist psychology and French phenomenology while extending Georges Gusdorf's language theory to writing. Baumlin regards Corder's heroic phase as a vernacular "West Texas elaboration upon French phenomenology" and his tragic phase as a "questioning and critique of the same." Baumlin argues that despite the influence of Gusdorf, Corder's "multiple rhetorics, adequation of rhetoric and psychology, and . . . theory of generative ethos" evince noteworthy originality. Baumlin further holds that Corderian rhetoric can contribute to the reexamination of social constructivism, which continues to dominate contemporary composition theory but which "has more or less completed itself and stands in need of correction."

Keith D. Miller in "Jim Corder's Radical, Feminist Rhetoric" contends that without saying so, Corder generates a systematic argument by indirection that undermines standard forms of academic persuasion. According to Miller, Corder enacts a strong but implicit logic that yokes anecdotes from rural West Texas to analyses of eighteenth-century British poetry to arguments about the theory and practice of teaching writing. Usually refusing to hand answers to readers, Corder prefers to invite them to join him in puzzling over problems. In doing so, he enacts what Miller calls "a radical, pioneering, subversive, and feminist rhetoric." W. Ross Winterowd in "The *Uses of Rhetoric*" analyzes the importance of Corder's *Uses of Rhetoric* and the reasons that it failed to reach the readers it deserved. Winterowd finds Corder "in a double bind, writing for an audience that needed basic education, but advancing an argument foreign to that audience's values and beyond what a reader with only the 'basics' could follow." Contemplating Corder's irascible oeuvre that folds classical rhetoric into expressivist discourse, Wendy Bishop in "Preaching What He Practices: Jim Corder's Irascible and Articulate Oeuvre" responds to and interprets several of his more significant essays, including "Argument as Emergence, Rhetoric as Love," "Notes on a Rhetoric of Regret," and "At Last Report I Was Still Here."

She notices his rather insistent use of classical schemes and tropes in essays that appear decidedly more informal than they are. Pat C. Hoy II, in "A Writer's Haunting Presence," gently probes and explores Corder's *Chronicle of a Small Town, Hunting Lieutenant Chadbourne*, and (especially) *Yonder*. Hoy quietly champions these texts, both as nonfiction and as rhetoric. By contrast, in "Finding Jim's Voice: A Problem in Ethos and Personal Identity," George E. Yoos weighs *Yonder* and finds it lacking the authenticity that Corder obviously sought. Not only does *Yonder* fail as a memoir, Yoos argues, it also stands as an epistemic failure and a moral failure. Unhappy with Corder's literary inclinations, Yoos also complains that unfortunately, Corder's scholarly career yields "the net effect of introducing tacitly literary standards and not cognitive standards into our perception of good writing."

In Part III, Parallels, Extensions, and Applications, contributors investigate and interrogate various trajectories, vectors, possibilities, and extensions of Corder's rhetoric. Responding to the recent wave of incivility in both private and public spaces, Theresa Enos in "A Call for Comity" considers observations and theories about comity from figures as diverse as Aristotle, Thucydides, Erasmus, Machiavelli, and Samuel Johnson. She considers Corder's advocacy of gentle persuasion while also engaging the rhetorical conceptions of Richard Weaver, Kenneth Burke, Jürgen Habermas, Stephen Carter, Deborah Tannen, and Richard Enos. Richard E. Young in "Toward an Adequate Pedagogy for Rhetorical Argumentation: A Case Study in Invention" tells about a recent class he taught in which he asked students to consider a case study involving the ethical issue of suicide. Concluding that the course failed, Young analyzes the failure and wonders about possibilities for transforming the course so that rhetoric becomes investigation as students reject dialogue-as-debate from fixed positions in favor of dialogue-as-discussion. Exploring the process of resolving conflicts, Richard Lloyd-Jones in "Rhetoric and Conflict Resolution" reflects on his wife's sixteen years of experience in the Iowa legislature and her role in founding the Iowa Peace Institute, which develops techniques for settling disputes and assists squabbling governmental bodies. He explores and reconceives the role of rhetoric in conflict resolution. Elizabeth Ervin in "Rhetoricians at War and Peace" explores rhetoric as an alternative to war—as has often been claimed—or whether war is simply an extreme expression of rhetoric. She considers Corder's and Wayne Booth's accounts of their stints in the US military and Lad Tobin's narrative about evading military service during the Vietnam War. She contemplates the relation between these scholars' discussions of their combat inexperience and their conceptions of rhetoric.

In Part IV, Theoretical, Pedagogical, and Institutional Issues, contributors offer future direction for some current issues. Defending expressivist rhetoric in "Bringing Over Yonder Over Here: A Personal Look at Expressivist Rhetoric as Ideological Action," Tilly Warnock explores her evolving conception of her childhood in Georgia and links her process of conceptualizing her early years to a consideration of rhetorical theory—a move, she claims, that Corder authorizes her to make. In "A More Spacious Model of Writing and Literacy," Peter Elbow explores some of the greatly varied, widely practiced forms of nonacademic, nonprofessional writing that scholars usually overlook, undervalue, and undertheorize. He offers a number of observations about these forms, observations that he modestly calls "fragments." John Warnock in "Weaving a Way Home: Composing a Personal Geography" considers the precarious, ill-defined position of geographical nonfiction and rhetoric in the university. Drawing on years of experience (and pondering a chapter in *Yonder*), he probes scholars' resistance to personal and regional rhetorics and argues that the academy should instead prize such discourses. Douglas Hesse in "Who Owns Creative Nonfiction?" ponders the institutional vagaries involved in teaching and administering courses in "creative nonfiction," which enjoys what he calls an uneasy relationship with the academy and which might belong either to rhetoric and composition studies or to creative writing. Hesse ponders which administrative domain is better suited to design and staff such classes and investigates writing curricula that seek to propagate inclusive rhetorics.

The contributors to this volume, in the spirit of Jim Corder's unfinished work, absorb, probe, stretch, redefine, and interrogate classical, modern, and postmodern rhetorics—and challenge their limitations. It is our hope that the essays can bolster our attempts—and generate new attempts—to develop postprocess composition theories and pedagogies and post-postmodern rhetorics.

We now turn to the work of Jim Corder, which he always conceived as "unfinished." In 1958, after completing his PhD at the University of Oklahoma, Corder began teaching at Texas Christian University, where he remained on the faculty for more than three decades. During the 1970s he joined a band of mutually supportive scholars in the Conference on College Composition and Communication (CCCC). The most obvious and most significant contribution of these professors—many of whom enjoyed highly productive careers over several decades—was to develop rhetoric and composition into an important, rapidly emerging field.

Corder helped build the discipline by publishing scores of essays, including many in *College English, College Composition and Communication*, and other major journals. He wrote one academic book, *Uses of Rhetoric*, and authored and edited numerous textbooks. A popular figure at CCCC gatherings, especially during the 1980s, he appeared on panels with such luminaries as Jim Berlin, Wayne Booth, Joseph Comprone, W. Ross Winterowd, and Richard Young. Beginning in the 1970s, Corder joined Winterowd in advocating the study of nonfiction literary prose, which departments of English characteristically devalued.

Spurred by Edward Corbett's *Classical Rhetoric for the Modern Student*, Corder became fascinated with ancient Greek and Roman rhetoric; developed a friendship with Corbett; and joined Corbett, Frank D'Angelo, James Kinneavy, Winifred Horner, and others in CCCC in revitalizing interest in Aristotle and Cicero. Prompted by Kenneth Burke, Corder stretched classical rhetoric to cover all forms of discourse—a project he began in 1971 when he wrote *Uses of Rhetoric*. Throughout his career he generated postclassical investigations, expansions, and redefinitions of invention, structure, style, memory, ethos, and other keystone conceptions of Aristotelian and Ciceronian rhetoric. During his midcareer and later, he explored postmodern and (arguably) feminist and post-postmodern perspectives on the same durable, yet (in his hands) elastic conceptions. At times he oscillates between two positions, attempting to welcome both but sometimes finding them opposed and pressing viselike against him.

Unlike Corbett, Corder combined his interest in classical rhetoric with a desire to add an expressivist dimension to scholarly writing. Beginning in the 1970s, he fused the academic and the personal by merging theories and scholarship about rhetoric and composition studies with narratives about an obscure corner of rural West Texas where he grew up during the Great Depression. He continued to blend personal and scholarly writing (sometimes leavening his essays with self-deprecating humor) during the early and middle 1980s, when composition was often viewed as a species of social science—a conception utterly foreign to him. After becoming an administrator during the early 1980s, he wrote—sometimes philosophically, sometimes humorously—about how university officials juggle dilemmas.

Although mixing expressivist (or autobiographical) and scholarly writing was highly unusual during most of Corder's career, Mike Rose, Victor Villanueva, Wendy Bishop, and others have greatly popularized the practice, which is now, in some quarters, de rigeur.

In "Argument as Emergence, Rhetoric as Love" and many other essays during the 1970s and 1980s, Corder advocates and enacts a dialogic rhetoric of mutual exploration that he hopes will replace agonistic rhetoric based on inflexible, predetermined positions. Beginning in 1989, he sacrificed part of his CCCC readership while producing works that are both personal and unclassifiable. Three of these idiosyncratic volumes—"Rhetoric, Remnants, and Regrets"; "Places in the Mind"; and "Scrapbook"—remain unpublished. Four of them—*Lost in West Texas, Chronicle of a Small Town, Yonder,* and *Hunting Lieutenant Chadbourne*—appeared before his death in 1998.

Like his other works (only more so), *Lost in West Texas, Chronicle of a Small Town, Yonder,* and *Hunting Lieutenant Chadbourne* prove difficult to summarize. In them Corder engages European postmodernism (usually implicitly) as he reinscribes the history of Jayton, Texas (and his own experience), while pondering the flawed nature of memory and what he terms his "disappearing" self. He wonders whether memory is as untrustworthy as it is indispensable and whether knowing the past is both necessary and impossible.

In later essays—including "Hunting Lieutenant Chadbourne" and "Notes on a Rhetoric of Regret"—Corder continues to engage postmodernism, which he could neither swallow nor escape but which, in "At Last Report I Was Still Here," he eventually appears to circumvent.

Despite Corder's many national publications, his writing has never been seriously assessed. No one has wondered in print whether "Corderian issues" and "Corderian rhetoric(s)" exist or what Corder's work might contribute to feminist rhetoric, postprocess pedagogy, or post-postmodern epistemology. While all the chapters in this book are unfinished and fragmentary, we hope that they can initiate such an assessment.

Keith D. Miller
Arizona State University

Theresa Enos
University of Arizona

I

Historical Context—Rhetoric and Composition Studies

1

The Spaciousness of Rhetoric

Janice Lauer
Purdue University

ℜThe title of this volume fits this introductory chapter because it states so well the scholarly climate and attitude of the 1960s and 1970s as rhetoric and composition was forming itself into a disciplinary field within the academy. In an earlier essay, "Dappled Discipline," I argued that composition studies (read rhetoric and composition) had already acquired the features of a discipline including special phenomena to study, modes of inquiry, a history of development, theoretical ancestors and assumptions, an evolving body of knowledge, and epistemic courts to grant status to new knowledge. It also had a department home, a ritual of academic preparation, and scholarly organizations and journals. That essay also related the field of rhetoric and composition to the teaching of writing but did not consider them synonymous. Today the nature of the field and its disciplinarity continue to be discussed and debated.

In this chapter I focus on the expansive 1960s and 1970s when people began to publish scholarship on writing and to speak of establishing an academic rhetorical field within English studies. Gradually during this period, courses, seminars, and finally graduate programs in rhetoric and composition emerged, thereby helping to construct a disciplinary domain. These startling developments required a lot of space for library expansion, intellectual territory, offices and classrooms, conferences, and tolerant, even exuberant, openness among those beginning to work in this area.

Many today are asking how this happened. Numerous explanations are being proffered by those constructing textual archives or genealogies. Ancestors are being extricated from earlier periods. One feature of the early 1960s that remains uncontested by those of us who were there is the lack of academic space for the study and research of written discourse (beyond literature). Those in English departments who became interested in the rhetoric of written discourse were considered odd. Their scholarship had no academic status, merited no tenure or promotion, and was relegated to the category of pedagogy and service in which the teaching of composition had long been situated. How did this status change? This chapter offers a synoptic remembrance of some of the motivating issues and factors that circulated during this period, contributing to the early work in rhetoric and composition that became a matrix for the field. Recalling the issues and alliances in the 1960s is important as a context for understanding the spacious tone of the 1970s.

MULTIDISCIPLINARY INFLUENCES

One factor was a cluster of multidisciplinary works in the 1950s and 1960s that entered the consciousness of those of us who became attuned to rhetoric: Kenneth Burke's rhetoric and grammar of motives, Lev Vygotsky's work on thought and language, Stephen Toulmin's arguments for informal reasoning, Michael Polanyi's notions of personal knowledge, Chaïm Perelman and Madame Lucie Olbrechts-Tyteca's *New Rhetoric*, Kenneth Pike's tagmemic linguistics, Craig La Driére's argument for a rhetorical kind of thinking, studies of creativity and heuristics in psychology, Walter Ong's study of Peter Ramus and the banishment of invention from rhetoric, and Daniel Fogerty's *Roots for a New Rhetoric*. The ideas and arguments of these works opened spaces for investigation and stimulated issues for inquiry among people interested in writing. Most of these works questioned prevailing views of the relationship between thought and discourse and of the nature and sources of reasoning. Some provided new analyses of the initiation and invention of arguments. Others, such as Pike's, challenged transformational grammar, offered a culturally grounded discourse theory of language use.

ALLIANCES IN THE 1960S

During the course of the 1950s and 1960s at the Conference on College Composition and Communication (CCCC) and in the journals, accounts of dissatisfaction with teaching writing increased. On a personal note, I had begun teaching in 1954 and by the 1960s had become frustrated with the options available for teaching composition. My education and preparation as a teacher did not include any help with teaching writing; everything I tried was ad hoc. At the college level, I was discontented with using case books and collections of essays organized by modes and with teaching grammar, linguistics, and logic in the composition course. This attitude was widespread. In 1963 Albert Kitzhaber published *Themes, Theories, and Therapy: The Teaching of Writing in College*, in which he surveyed freshman composition courses, lamenting the lack of academic preparation in rhetorical theory for teaching composition and the plethora of diverse untheorized textbooks.

In the multidisciplinary discourses mentioned above, we found starting points for research about composition. Our new interests, however, might have spun in endless space, unconnected anomalies outside the purview of literary or linguistic studies, had we not begun to form some scholarly alliances. The one I'm most familiar with is the Rhetoric Society of America, whose first meeting took place in 1968 at the CCCC in Minneapolis. As I have recounted elsewhere, this group drew together people in English, communication, philosophy, and other fields to share their work in rhetorical studies. We continued to meet at the CCCC. An invitation for membership appended to the first issue of *The Rhetoric Society Newsletter* explained the purpose of the society: to (a) "promote communication among those who are concerned with rhetoric," (b) "disseminate knowledge of rhetoric and the powers of rhetoric to those who have been previously unaware of it," and (c) "encourag[e] direct implementation of experimentation in, and research into the implementations of rhetoric in composition, speech, and communication courses" (1-2). The membership list published in the first volume of the *Newsletter* included such people in English as Wayne Booth, Ed Corbett, Jim Corder, Frank D'Angelo, Paul Doherty, Walker Gibson, Robert Gorrell, William Irmsher, James Kinneavy, Albert Kitzhaber, Richard Larson, Janice Lauer, Richard Ohmann, Gordon Rohman, Gary Tate, Winston Weathers, Ross Winterowd, and Richard Young.

After our annual RSA meetings at the CCCC, a small group of us went to dinner and talked about a possible field of rhetoric within English studies. The group included Ed Corbett, Ross and Norma Winterowd, Richard Young, George and Mary Yoos, and me. Members of this group later initiated institutional formations that ultimately played an important role in the development of disciplinarity, offering courses, national seminars, and, eventually in the 1970s and early 1980s, doctoral programs. Was there competition among these seminars and programs? Were we jostling for space? Not at all. The first national Rhetoric Seminar that Ross Winterowd and I directed in 1976 brought together an unanticipated sixty-five participants for two weeks of intense study of rhetoric and composition theory, interacting with theorists in the field. By the end of the decade, so many people were interested in pursuing rhetorical studies of written discourse that we appreciated each others' efforts to service these needs. As new studies of written discourse emerged, they were welcomed by a growing cadre of people identifying with rhetoric and composition. Critical dialogue such as the exchange between Ann Berthoff and me in *College Composition and Communication* enabled theories and ideas to be qualified and refined.

These exchanges and alliances formed what might be called a critical mass, a group of people across the country with roots in composition and branches in rhetoric. Within this group were journal editors: William Irmsher and Ed Corbett with *College Composition and Communication,* Richard Ohmann with *College English,* and George Yoos with *The Rhetoric Society Newsletter*. Although journal space was limited, editors managed to find room for much of the early scholarship of rhetoric and composition.

SCHOLARLY ISSUES

In the 1960s three issues caught the attention of people associated with composition instruction: the lack of attention to invention and audience, the impressionistic bases for principles of style, and the dominance of the four modes of discourse. Three such broad areas for inquiry would be hard to identify, yet scholars were not daunted but rather excited by the vast possibilities for investigation. It was as if, rhetorically famished, we had stumbled onto a rich landscape lying fallow for centuries.

Invention and Audience

Studies of invention by composition theorists arose to meet several issues circulating at that time: the historical loss of invention, discussed by Ong and others; dialogues over the question of rhetoric as epistemic, initiated by Robert L. Scott's 1967 essay, "On Viewing Rhetoric as Epistemic"; Dudley Bailey's 1964 "A Plea for a Modern Set of Topoi"; and debates about the nature of arguments and sources of arguments and persuasion, found in Toulmin's 1969 *The Uses of Argument* and Perelman and Olbrechts-Tyteca's 1969 *The New Rhetoric*. After centuries of rhetoric's relegation to delivering the knowledge constructed by philosophy or science, scholars in communication and the developing field of rhetoric and composition began to research rhetoric's role in knowledge construction and to create inventional arts to guide that knowledge-making process.

From its beginning, the investigation of invention was inevitably entwined with conceptions of writing as a process. In 1964 Gordon Rohman and Albert Wlecke's *Prewriting: The Construction and Application of Models for Concept Formation in Writing* articulated the issue at stake:

> A fundamental misconception which undermines so many of our best efforts in teaching writing: If we train students how to recognize an example of good prose ("the rhetoric of the finished word"), we have given them a basis on which to build their own writing abilities. All we have done, in fact, is to give them standards to judge the goodness or badness of their finished effort. *We haven't taught them how to make that effort.* (106)

Although their term *prewriting* later came to be misunderstood as nondiscursive thinking, their study and Rohman's subsequent articles addressed several significant aspects of invention: They positioned invention as a crucial part of a process, introduced new inventional strategies, conducted research on invention, examined invention by drawing on theories in other fields, used the term *rhetoric* as synonymous with discourse, and introduced invention into pedagogy.

In 1965 Edward Corbett's *Classical Rhetoric for the Modern Student* presented versions of the classical inventional arts of status, topoi, and the enthymeme, providing a rhetorical education for composition instructors and students. Also in 1965 Ross Winterowd's *Rhetoric and Writing* included sections on invention that discussed Aristotle's four requisites for addressing an audience, Cicero's stress on learning, Kenneth Burke's principle of courtship, and Richard Weaver's three classical sources of argument. In 1968 Winterowd, in *Rhetoric: A Synthesis*, aimed to establish a "new basis for a new kind of rhetoric" (v-vi), drawing on the work of Burke and Francis Christensen. Also in 1967 in my dissertation at the University of Michigan, "Invention in Contemporary Rhetoric: Heuristic Procedures," I analyzed rhetorical theories of the day (for example, neo-Aristotelianism, Burke's theories, I. A. Richards' work, general semantics, tagmemic rhetoric, Toulmin's model, and others), searching for manifestations of inventional theories or practices and reading incipient theories of invention through the lens of emerging scholarship on heuristic thinking in creativity studies. The concept of heuristics as an alternative to logical thinking offered the field another way of theorizing composition as epistemic. In 1969 Janet Emig finished her dissertation at Harvard, later published as *The Composing Processes of Twelfth Graders*. Not only did her work strengthen an interest in the composing process, but also through a case study (a new research mode for studying writing), she differentiated, observed, and described inventional acts such as prewriting, planning, and starting.

In 1970 Richard Young and Alton Becker published their first theoretical essay on tagmemic rhetoric, followed by their book *Rhetoric: Discovery and Change*. This new modern rhetorical theory, based on Kenneth Pike's tagmemic linguistics, addressed the entire composing process from invention, through audience and organization, to style. Their theory of invention, positioned within the framework of an inquiry process, offered a strategy for initiating the writing process, a new epistemological exploratory guide, and guidelines for verifying new understandings.

As one would expect, the revival of rhetoric also brought concerns about the lack of attention to audiences for written discourse. Earlier work suggested possible audience theories such as Burke's notion of identification rather than persuasion and Perelman and Olbrecht-Tyteca's audience-based sources of argument. Within English studies, catalysts in this direction were Wayne Booth's 1963 "The Rhetorical Stance" and later tagmemic rhetoric's Rogerian audience theory.

It is interesting to note that these studies of invention and audience already bore the marks of this new field because they not only employed multimodality, multidisciplinary sources, and theoretical arguments, but they also worked out the implications for composition instruction.

Studies of Style, Grammar, Form, Voice, and Rhetorical Criticism

Unlike the work on invention, which addressed a neglected aspect of written discourse, studies of grammar, style, voice, coherence, and form in the 1960s sought to add rigor and theory to existing emphases in composition instruction. In 1963 Richard Braddock, Richard Lloyd-Jones, and L. Schoer's *Research in Written Composition* revealed the preoccupation of researchers with error counts and problems with grammar instruction and claimed that "The teaching of formal grammar has a negligible or, because it usually displaces some instruction and practice in actual composition, even harmful effect on the improvement of writing" (37-38). In the 1960s Louis Milic published several works on style including "Theories of Style and Their Implications for the Teaching of Composition," "Against the Typology of Styles," and *A Quantitative Approach to the Style of Jonathan Swift*, which argued for a more rigorous rather than impressionistic analysis of style. In that same year, based on transformational grammar, Richard Ohmann's "Literature as Sentences" explained how authors' syntactic fluency revealed their habits of meaning and deep linguistic resources. In 1967 Francis Christensen challenged reigning discussions of the sentence that praised a periodic structure, arguing instead for the cumulative sentence and paragraph, based on analyses of current authors. Also influential were Walker Gibson's 1966 *Tough Sweet Stuffy*, which analyzed prose styles—demonstrating how every writer's choice dramatized a personality or voice in relation to a center of concern and the person being addressed—and his 1969 *Persona: A Style Study for Readers and Writers*, which demonstrated a research interest in style. Kenneth Macrorie's 1968 *Writing to Be Read* and 1970 *Telling Writing* questioned the traditional advice about matters of style and voice, battling against "Engfish" and creating reasoned descriptions and categories. Another prominent text was Winston Weathers and Otis Winchester's *Copy and Compose: A Guide to Prose Style* in 1969, which built on Winterowd's work on style and form in *Rhetoric: A Synthesis* and extended Corbett's focus on schemes and tropes in *Classical Rhetoric for the Modern Student*. A number of collections of essays on style were also published, including Glen Love and Michael Payne's *Contemporary Essays on Style: Rhetoric, Linguistics, and Criticism* and Louis Milic's *Stylists on Style*.

Rhetorical analysis was advocated by Corbett's 1966 pioneering essay, "A Method of Analyzing Prose Style with a Demonstration Analysis of Swift's 'A Modest Proposal'" and his *Rhetorical Analyses of Literary Works* in 1969, which included Jim Corder's 1967 essay in *PMLA*, "Rhetoric and Meaning in *Religio Laici*," which interpreted Dryden's poem as an example of a classical text and referred to a reconciliation between rhetoric and poetic.

These discussions of grammar, style, form, voice, and rhetorical analysis moved beyond New Criticism's worship of the well-wrought urn standing apart from author, audience, and context to considerations of the rhetorical situation with its readers and authorial voice in prose texts of all kinds. Here again, the scope of rhetorical studies expanded.

Theories of Discourse

A third space for scholarship in the 1960s opened with the discourse theory of James Moffett based on language development research and theories of symbolic hierarchies and abstraction. In his 1968 *Teaching the Universe of Discourse*, he constructed a curriculum integrating the language arts, offering two new bases for grouping discourse: the levels of abstraction between speaker and listener and between speaker and subject. In 1969 James Kinneavy extended this path, publishing "The Basic Aims of Discourse," his first essay on his theory of discourse, which anticipated his 1971 *A Theory of Discourse*. Basing his ideas on an extensive study of semiotic theory, Kinneavy argued for four aims of discourse—expressive, persuasive, referential, and literary—analyzing, exemplifying, and differentiating them from modes of discourse and demonstrating their equal importance in composition instruction.

Pedagogies

These broad areas of inquiry began to generate new pedagogies: process teaching that engaged instructors and students in such inventional acts as initiating writing with issues or questions, exploring using different guides, analyzing audiences, and framing judgments. Guiding these acts were many of the new heuristics. Although theoretical distinctions had been made between epistemological heuristics and those used to find arguments for claims already held, the textbook adaptations often described heuristics only as the latter. Further, inventional acts of initiation and exploration still often took place within modal frames, that is, writing a description, narration, exposition, or argument. In some cases the process became simplified into prewriting, writing, and rewriting, often misrepresenting "prewriting" as a set of mental acts and process as linear. Examples of widely used textbooks included the *Harbrace Handbook* and James McCrimmon's *Writing with a Purpose*, which began to change to reflect developments in the field.

Disciplinarity

Although much of the publication in this period was devoted to research in the above areas, a few essays addressed the question of disciplinarity and the state of the developing discipline itself. The early ones articulated definitions and frameworks for a rhetorical discipline. In 1965 Virginia Burke reported that her survey of *College English* and *College Composition and Communication* from 1950-1964 did not reveal any mention of the word *rhetoric*, but that in 1965, *rhetoric* was "a magic word" (3). She reported on a 1961 session at the CCCC, "Rhetoric—The

Neglected Art" (the first session to deal with rhetoric) in which the participants "had arrived at a useful distinction—that courses emphasizing rhetorical principles should focus on *all aspects of effective oral and written discourse* as contrasted with courses which are concerned primarily with grammar and mechanics of writing or with literature" (5; emphasis added). This description carved out a spacious site for rhetoric and composition indeed. But alongside this rhetorical resurgence, she lamented in 1965, there was still "chaos in the teaching of composition because . . . composition has lacked an informing discipline, without which no field can maintain its proper dimensions . . . or its very integrity" (5). In the 1965 *College Composition and Communication*, Robert Gorrell discussed the conclusions of a seminar convened by the CCCC to discuss the recent rhetorical phenomenon. (Participants were Wayne Booth, Virginia Burke, Francis Christensen, Edward Corbett, Robert Gorrell, Albert Kitzhaber, Richard Ohmann, James Squire, and Richard Young.) Gorrell reported that they had discussed the manifestations of rhetoric at the time and provided the definition of rhetoric that was generally understood by those beginning to link rhetoric to composition as "the theoretical study of discourse, including the relations between thought and language, the patterns of modern prose, and differences between speech and writing" (142). The seminar group agreed that

> rhetoric considered as practical advice about speaking and writing grows from comprehensive rhetorical theory, warning, however, that teaching rhetoric [in composition courses] must be neither a direct exposition of theory nor a collection of rules or warnings. It must rather attempt to describe the choices available to the writer, to explain the results or effects of different choices, and thereby give the writer a basis for choosing. (142)

They concluded that such a rhetorical theory must be "comprehensive enough to take account of all aspects of composition," and they called for "a course on the college level which concentrates on developing the art of discourse and that rhetoric is logically the central subject matter of this course" (143). They also saw the importance of graduate work in rhetorical theory. Such a charter was expansive. Only two years later in 1967, Martin Steinmann compiled a record of accumulating scholarly essays on rhetoric, invention, and style, titling it *New Rhetorics*.

The emerging field in the 1960s was broad, intertextual, and reflective of the personal and intellectual networks that had been building. It was not just a textual space but a haven where we could go to be understood, valued, and energized. The new discipline formed the context for the spacious 1970s in which theorists enlarged these discussions and exploded into many new areas.

THE 1970S

Invention, Audience, and Ethos

In the 1970s inventional studies continued to be published: for example, "Ann Berthoff's 1972 "From Problem-solving to a Theory of the Imagination" and her 1978 *Forming, Thinking, Writing: The Composing Imagination* as well as Lee Odell's 1973 "Piaget, Problem-Solving, and Freshman Composition." In 1975 Frank D'Angelo's *A Conceptual Theory of Rhetoric*, drawing on cognitive psychology, psycholinguistics, and classical and modern rhetoric, built a rhetorical theory based on abstract mental structures that determined discourse and provided a description of a writer's competence rather than performance. In the same year, Walter Ong contributed to audience theory in his essay, "The Writer's Audience is Always a Fiction," which posited readers constructed by writers in their texts. In the 1970s Linda Flower and John Hayes presented their first essays on cognitive process theory including "Problem-Solving Strategy and the Writing Process" and Flower's 1979 "Writer-Based Prose: A Cognitive Basis for Problems in Writing." Based on protocol analyses of composing processes, their model described a set of distinctive thinking processes orchestrated during composing, embedded in one another, and goal directed. Richard Larson illustrated how the problem-solving process informed many kinds of writing. In 1978 Jim Corder addressed ethos or ethical argument in "Varieties of Ethical Argument, With Some Account of the Significance of *Ethos* in the Teaching of Composition," stressing the importance of studying how a trustworthy ethos is developed in language and what qualities of character elicit accord. These scholars working on invention, audience, and ethos refined, qualified, extended, or sometimes challenged previous work, but their attitude continued to be, in my estimation, one of mutual appreciation.

Theories of Discourse

In 1975 in *Writing: Basic Modes of Organization,* written with John Cope and J. W. Campbell, Kinneavy developed his theory of modes—description, narration, classification, and evaluation—based on classical notions of status, the *progymnasmata*, and philosophical notions of static and dynamic. In that same year, James Britton, Tony Burgess, Nancy Martin, Alex McLeod, and Harold Rosen framed another set of discourse categories (expressive, transactional, poetic) based on the roles of participant and spectator in language use and on the work of language theorists such as Suzanne Langer, Edward Sapir, James Bruner, Lev Vygotsky, Roman Jacobson, and Dell Hymes. They conducted extensive research in the schools in England to determine which types of discourse and audiences were most frequently taught, finding that transactional writing, especially analogic and report, predominated as did the teacher as examiner. Like those in the 1960s, these discourse theories were motivated by issues in curricular emphasis, the stranglehold of the four modes of discourse (exposition, description, narration, and argumentation), the dominance of exposition, and a lack of defensible bases for discursive classifications.

Style, Form, and Rhetorical Analyses

The early 1970s also saw work in style turn toward interest in syntactic fluency. In 1973 Frank O'Hare published *Sentence Combining: Improving Student Writing without Formal Grammar Instruction*. In a school experiment, he demonstrated that the students improved their syntactic fluency through sentence combining without being taught formal grammar. This research stimulated widespread pedagogies and numerous other studies and books followed including, in 1979, Donald Daiker, Andrew Kerek, and Max Morenberg's *The Writer's Options: College Sentence Combining*. Other work on style included Louis Milic's "The Problem of Style," a discussion of stylistics as a more rigorous method using computers to analyze style; Richard Ohmann's arguments for the usefulness of speech-act theory for stylistic analysis in his 1971, "Speech, Action, and Style"; Winston Weather's analysis of alternative contemporary styles in "Grammars of Style: New Options in Composition," and Joseph Williams' analysis of style in "Defining Complexity." Studies of form at this time included Ross Winterowd's 1971 "Dispositio: The Concept of Form in Discourse" and his 1979 "The Grammar of Coherence," as well as Richard Larson's "Toward a Linear Rhetoric of the Essay." These and many other studies of style and form held a range of implications for teaching.

In this decade our field promulgated an important statement. Fortified with a bibliography of 129 entries of research and theory in sociolinguistics, rhetoric, and other fields, the Committee on the CCCC Language Statement began its work, which culminated in "Students' Right to Their Own Language" in 1974, challenging the requirement for a single American Standard English in student writing with its erasure of differences and affirming "the students' right to their own patterns and varieties of language—the dialects of their nurture or whatever dialects in which they find their own identity and style" (2).

In 1971 Jim Corder published *Uses of Rhetoric*, which featured rhetorical analyses of a wide range of discourse: advertisements, essays, poems, and composition textbooks. In 1976 he authored the bibliographic essay on this aspect of rhetoric in the Tate collection. Although he maintained that all analyses of writing, including items in the MLA bibliography, were rhetorical, he focused on those works that used the language of rhetoric, analyses of nonfiction prose, and works after 1965. He further clustered titles under those that offered bases for rhetorical criticism and those that provided guides for such criticism, including historical studies of rhetoric that provided guides for analyses.

Basic Writing and Writing Development

In response to the cultural situation of underprepared students populating college classrooms, a fourth avenue of inquiry began with the work of Mina Shaughnessy, whose 1978 *Errors and Expectations* focused on errors as an object for inquiry, using discourse analysis methods and interpretation. Adding to this line of scholarship in 1978 in "What We Know—and Don't Know—about Remedial Writing," Andrea Lunsford presented research from her dissertation on basic writing programs. In 1979 Sondra Perl's "The Composing Processes of Unskilled College Writers" used a case study to examine the processes of basic writers. In the 1970s Muriel Harris also addressed a number of aspects of basic writing: "Making the Writing Lab an Instructor's Resource Room"; "Individualized Diagnoses: Searching for Causes, Not Symptoms of Writing Deficiencies"; and "Contradictory Perceptions of Rules of Writing"; and in 1977 initiated the *Writing Lab Newsletter*. This work on remediation and basic writing branched out eventually into an ancillary field with its own organizations, conferences, and journals. Into this kaleidoscopic and expanding space during the 1970s, researchers in the United States and Canada also introduced studies of writing development. In this vein were Barry Kroll's 1977 dissertation, "Cognitive Egocentrism and Written Discourse" and Carl Bereiter's 1979 "Development in Writing."

Revision and Evaluation

In 1978 a new aspect of the composing process gained saliency with Don Murray's writings on revision as "seeing again"; Ellen Nold's work on surface and meaning-based revision; and Nancy Sommers' dissertation, "Revisions in the Composing Process: A Case Study of College Freshman and Experienced Adult Writers."

In the 1970s we also became preoccupied with theories of evaluation: holistic, analytic, and primary trait scoring. In 1974 Paul Diederich described analytic categories in his *Measuring Growth in English*. In 1977 Charles Cooper and Lee Odell edited a collection of essays, *Evaluating Writing: Describing Measuring Judging*, that drew together discussions of assessments then in place, including Richard Lloyd-Jones's essay on Primary Trait Scoring developed by the National Assessment for Educational Progress and Educational Testing Service's General Impression Scoring. In 1975 John Daly and M. Miller developed a measure to gauge writing apprehension, described in "Writing Apprehension as a Predictor of Message Intensity."

Pedagogies

New pedagogies gained followers during this decade. Textbooks began to include a variety of heuristics for invention—for example, some classical topics (used as an exploratory cluster rather than as discrete modes), versions of the tagmemic heuristic, and Burke's Pentad. David Harrington and others compiled these heuristics in a 1979 article, "A Critical Survey of Resources for Teaching Rhetorical Invention: A Review Essay." Prominent also were cognitive process methods described by Flower and Hayes in "Problem-Solving Strategies and the Writing Process," freewriting advanced in 1975 by Peter Elbow, and conferencing encouraged by Don Murray. Some of these approaches to teaching were represented in Timothy Donovan and Ben McClelland's collection of essays, *Eight Approaches to Teaching Composition*.

Disciplinarity

This decade also witnessed more explicit efforts toward disciplinarity: early theoretical discussions of the nature of rhetoric and composition as a scholarly field, the appearance of national conferences and seminars, and the founding of doctoral programs. In his 1975 important book, *Contemporary Rhetoric: A Conceptual Background with Readings*, Ross Winterowd gathered and contextualized some of the key studies on invention, form, and style as a textual record of the developing field of rhetoric and composition. In 1976 Gary Tate provided the field with a collection of essays, *Teaching Composition: Ten Bibliographic Essays*, an impressive list of scholarship in rhetoric and composition.

In addition to the CCCC and the Rhetoric Society meetings, national seminars and conferences gave forums to the developing field of rhetoric and composition. In 1976 the Rhetoric Society of America held an invitational conference at the University of Wyoming, whose purpose was to plan multidisciplinary national seminars. At this invitational conference, eleven of us (Frank D'Angelo, Dorothy Guinn, E. D. Hirsch, Janice Lauer, Michael Leonard, Ellen Nold, Patricia Sullivan, John Warnock, Ross Winterowd, George Yoos, and Richard Young) exchanged ideas and work in progress. This small conference seemed to me at the time a microcosm of the collegiality of the field. Two other conferences in the decade included the Wyoming Conference held annually and, beginning in 1976, the Rhetoric Seminar, a two-week study of rhetoric and composition theory, held for five years at the University of Detroit and for eight years at Purdue University. Each summer at this seminar, fifty to sixty participants had the opportunity to read and discuss the emerging theory and research in the field with eight scholars authoring that work. In retrospect, these conferences and seminars were community-building experiences, educating people in the literature of composition and rhetoric and creating national networks and alliances.

By the end of the 1970s, several doctoral programs in rhetoric and composition had been established, including the University of Southern California's program and fifteen other programs listed in *Rhetoric Review*'s 2000 survey of doctoral programs. With the opportunity to educate new members through these doctoral programs, disciplinarity was solidified. Rhetoric and composition as a scholarly field found an academic home. Reports of undergraduate rhetoric programs were also appearing. In 1973 the University of Tulsa outlined its new undergraduate program in rhetoric and writing, which included courses in linguistics and rhetorical theory and emphasized rhetoric in action. In 1975 George Tade, Gary Tate, and Jim Corder proposed a university-wide curriculum based on invention, disposition, and style.

As the decade closed, the Canadian Council of Teachers of English Conference Learning to Write (The Ottawa Conference) brought together in 1979 scholars from the United States, Canada, British Isles, and Australia. In the volume published subsequently, *Reinventing the Rhetorical Tradition*, Janet Emig in "The Tacit Tradition: The Inevitability of a Multi-Disciplinary Approach to Writing Research" wrote an apologia for rhetoric and composition as a discipline, arguing that the field had a multidisciplinary intellectual tradition, modes of inquiry, and a tacit tradition comprised of scholars who, despite coming from different vantages, affirmed the importance of the nature of learning and language, especially written language, and provided research into writing (10). The editors of the collection, Ian Pringle and Aviva Freedman, stated that "To many who were present . . . it seemed that the conference served as a culmination of all that had been achieved in the study of rhetoric since the beginning of the recent resurgence of interest in the discipline" (173).

Jim Corder's work in these decades also exemplifies this history of an emerging discipline, beginning with his early membership in the Rhetoric Society and his 1971 *Uses of Rhetoric* and continuing on to his publications on ethics and rhetorical analysis. Throughout his writing and curriculum design, he positioned rhetoric at the center of composition.

What legacies have Jim Corder and the other theorists in these decades left the field? Jim's abiding interest in ethos and ethical argument stands as a harbinger of the intense discussion today on ethics and rhetoric. From this period the field also inherits the restless spirit of inquiry of the theorists. Just as their quest and curiosity for knowledge about rhetoric and written discourse could not be quenched by the initial lack of disciplinary structures and academic support, so too the field now continues to follow its leads into new avenues of scholarship, at times building on former insights and other times taking new theoretical directions. Just as rhetoric was seen then as contextualized discourse and as a core of their investigation and teaching, so too do postmodern concepts of situatedness and constructedness drive the field to raise important questions about agency, writer and reader positions, and difference. As early interest in invention stimulated a range of theories about writing processes and heuristic procedures, so too does the field now study new critical practices and the construction of knowledge in workplaces, disciplines, and public spheres. Just as discourse theorists challenged the hegemony of expository writing and the four modes of discourse, so too the field now examines multiple genres in different discourse communities and cultures. These legacies inhabit the disciplinary space in which the field now investigates and publishes. But it remains to be seen whether the spirit of collegiality will continue to characterize the field, engendering respect both for this earlier work and for the current extensive range of scholarship and teaching, regardless of mode of inquiry or ideology.

In these remarkable two decades, the space for rhetoric and composition grew exponentially from one peopled with isolated individuals and free-floating rhetorical interests to situated scholarship; from preoccupations with grammar to exploding research on myriad aspects of writing; from a small number of rhetorical devotees to dozens of theorists, writing instructors, and graduate students focused on rhetorical issues. The mood was enthusiastic; the prospects for significant and rewarded scholarship seemed unlimited. The spirit of collegiality flourished. The spaciousness of rhetoric prevailed.

WORKS CITED

Bailey, Dudley. "A Plea for a Modern Set of Topoi." *College English* 26 (1964): 111-17.

Bereiter, Carl. "Development in Writing." *Testing, Teaching, and Learning. Report of a Conference on Research on Testing.* Washington, DC: National Institute of Education, 1979.

Berthoff, Ann. *Forming, Thinking, Writing: The Composing Imagination.* Rochelle Park, NJ: Hayden, 1978.

—. "From Problem-Solving to a Theory of Imagination." *College English* 33 (1972): 636-51.

—. "The Problem of Problem Solving." *College Composition and Communication* 22 (1971): 237-42.

—. "Response to Janice Lauer." *College Composition and Communication* 23 (1972): 414-15.

Booth, Wayne. "The Rhetorical Stance." *College Composition and Communication* 14 (1963): 139-45.

Braddock, Richard, Richard Lloyd-Jones, and L. Schoer. *Research in Written Composition.* Champaign, IL: NCTE, 1963.

Britton, James, et al. *The Development of Writing Abilities (11-18).* London: Macmillan, 1975.

Burke, Kenneth. *A Grammar of Motives.* Berkeley: U of California P, 1945.

—. *A Rhetoric of Motives.* Berkeley: U of California P, 1950.

Burke, Virginia. "The Composition-Rhetoric Pyramid." *College Composition and Communication* 16 (1965): 3-7.

Christensen, Francis. "A Generative Rhetoric of the Paragraph." *College Composition and Communication* 16 (1965): 144-56.

—. "A Generative Rhetoric of the Sentence." *College Composition and Communication* 14 (1963): 155-61.

—. *Notes Toward a New Rhetoric*. New York: Harper, 1967.

Cooper, Charles, and Lee Odell, eds. *Evaluating Writing: Describing Measuring Judging*. Urbana, IL: NCTE, 1977.

Corbett, Edward. "A Method of Analyzing Prose Style with a Demonstration Analysis of Swift's 'A Modest Proposal.'" *Reflections on High School English*. Ed. Gary Tate. Tulsa, OK: U of Tulsa, 1966. 312-30.

—. *Classical Rhetoric for the Modern Student*. New York: Oxford UP, 1965.

—. *Rhetorical Analyses of Literary Works*. New York: Oxford UP, 1969.

Corder, Jim W. "Rhetorical Analyses of Writing." *Teaching Composition: Ten Bibliographic Essays*. Ed. Gary Tate. Fort Worth: Texas Christian UP, 1976. 223-40.

—. "Rhetoric and Meaning in *Religio Laici*." *PMLA* 82 (1967): 245-50.

—. *Uses of Rhetoric*. New York: Lippincott, 1971.

—. "Varieties of Ethical Argument, With Some Account of the Significance of *Ethos* in the Teaching of Composition." *Freshman English News* 6 (1978): 1-23.

Daiker, Donald, Andrew Derek, and Max Morenberg, eds. *The Writer's Options: College Sentence Combining*. New York: Harper, 1979.

Daly, John, and M. Miller. "Writing Apprehension as a Predictor of Message Intensity." *Journal of Psychology* 89 (1975): 175-77.

D'Angelo, Frank. *A Conceptual Theory of Rhetoric*. Englewood Cliffs, NJ: Winthrop, 1975.

Diederich, Paul. *Measuring Growth in English*. Urbana, IL: NCTE, 1974.

"Doctoral Programs in Rhetoric and Composition." *Rhetoric Review* 18 (Spring 2000): 244-373.

Donovan, Timothy, and Ben McClelland, eds. *Eight Approaches to Teaching Composition*. Urbana, IL: NCTE, 1980.

Elbow, Peter. *Writing Without Teachers*. New York: Oxford UP, 1973.

Emig, Janet. *The Composing Processes of Twelfth Graders*. Urbana, IL: NCTE, 1971.

—. "The Tacit Tradition: The Inevitability of a Multidisciplinary Approach to Writing Research." *Reinventing the Rhetorical Tradition*. Ed. Ian Pringle and Aviva Freedman. Akron, OH: L & S, 1980. 9-17.

Flower, Linda. "Writer-Based Prose: A Cognitive Basis for Problems in Writing." *College English* 41 (1979): 19-37.

Flower, Linda, and John Hayes. "Problem-Solving Strategies and the Writing Process." *College English* 39 (1977): 449-61.

Fogarty, Daniel. *Roots for a New Rhetoric*. New York: Columbia UP, 1959.

Freedman, Aviva, and Ian Pringle, eds. *Reinventing the Rhetorical Tradition*. Akron, OH: L & S, 1980.

Gibson, Walker. *Persona: A Style Study for Readers and Writers*. New York: Random, 1969.

—. *Tough Sweet Stuffy*. Bloomington: Indiana UP, 1966.

Gorrell, Robert. "Very Like Whale—A Report on Rhetoric." *College Composition and Communication* 16 (1965): 138-43.

Harrington, David V., Philip M. Keith, Charles W. Kneupper, Janie A. Tripp, and William F. Woods. "A Critical Survey of Resources for Teaching Rhetorical Invention: A Review Essay." *College English* 40 (1979): 641-61.

Harris, Muriel. "Contradictory Perceptions of Rules of Writing." *College Composition and Communication* 30 (1979): 218-20.

—. Individualized Diagnosis: Searching for Causes, Not Symptoms of Writing Deficiencies." *College English* 40 (1978): 64-69.

—. "Making the Writing Lab an Instructor's Resource Room." *College Composition and Communication* 28 (1977): 376-78.

Kinneavy, James. "The Basic Aims of Discourse." *College Composition and Communication* 20 (1969): 297-304.

—. *A Theory of Discourse*. New York: Norton, 1971.

Kinneavy, James, et al. *Writing: Basic Modes of Organization*. Dubuque, IA: Kendall/Hunt, 1976.

Kitzhaber, Albert. *Themes, Theory, and Therapy: The Teaching of Writing in College*. New York: McGraw, 1963.

Kroll, Barry. "Cognitive Egocentrism and Written Discourse." Diss. U of Michigan, 1977.

La Driére, Craig. "Rhetoric and 'Merely Verbal' Art." *English Institute Essays*. Ed. D. A. Robertson. New York: Columbia UP, 1949. 134-52.

Larson, Richard. "Problem-Solving, Composing, and Liberal Education." *College English* 33 (1972): 623-35.

—. "Toward a Linear Rhetoric of the Essay." *College Composition and Communication* 22 (1971): 140-46.

Lauer, Janice M. "Composition Studies: Dappled Discipline." *Rhetoric Review* 3 (1984): 20-29.

—. "Invention in Contemporary Rhetoric: Heuristic Procedures." Diss. U of Michigan, 1967.

—. "Response to Ann E. Berthoff." *College Composition and Communication* 23 (1972): 208-10.

Lloyd-Jones, Richard. "Primary Trait Scoring." *Evaluating Writing: Describing Measuring Judging*. Ed. Charles Cooper and Lee Odell. Urbana, IL: NCTE, 1977. 33-66.

Love, Glen, and Michael Payne. *Contemporary Essays on Style: Rhetoric, Linguistics, and Criticism*. Glenview, IL: Scott, Foresman, 1969.

Lunsford, Andrea. "Cognitive Development and the Basic Writer." *College English* 41 (1979): 38-46.

—. "What We Know—and Don't Know—About Remedial Writing." *College Composition and Communication* 29 (1978): 47-52.

McCrimmon, James. *Writing with a Purpose*. 4th ed. New York: Houghton, 1967.

Macrorie, Ken. *Telling Writing*. New York: Hayden, 1970.

—. *Writing to be Read*. New York: Hayden, 1968.

Milic, Louis. "Against the Typology of Styles." *Essays on the Language of Literature.* Ed. Seymour Chatman and Samuel Levin. Boston: Houghton, 1967. 442-50.

—. *A Quantitative Approach to the Style of Jonathan Swift.* The Hague: Mouton, 1967.

—. "The Problem of Style." W. Ross Winterowd. *Contemporary Rhetoric: A Conceptual Background with Readings.* New York: Harcourt, 1975. 271-95.

—. *Stylists on Style: A Handbook with Selections for Analysis.* New York: Scribner's, 1969.

—. "Theories of Style and Their Implications for the Teaching of Composition." *College Composition and Communication* 16 (1965): 66-69, 126.

Moffett, James. *Teaching the Universe of Discourse.* Boston: Houghton, 1968.

Murray, Donald. "Internal Revision: A Process of Discovery." *Research on Composing: Points of Departure.* Ed. Charles Cooper and Lee Odell. Urbana, IL: NCTE, 1978. 85-103.

—. "The Listening Eye: Reflections on Writing Conferences." *College English* 41 (1979):13-18.

Nold, Ellen. "Alternatives to Madhatterism." *Linguistics, Stylistics, and the Teaching of Composition.* Ed. Donald McQuade. Akron, OH: L & S, 1979. 103-17.

O'Hare, Frank. *Sentence Combining: Improving Student Writing without Formal Grammar.* Urbana, IL: NCTE, 1973. 19-34.

Odell, Lee. "Piaget, Problem-Solving, and Freshman Composition." *College Composition and Communication* 24 (1973): 36-42.

Ohmann, Richard. "Generative Grammars and the Concept of Literary Style." *Word* 20 (1964): 423-39.

—. "Literature as Sentences." *College English* 27.4 (1966): 261-67.

—. "Speech, Action, and Style." Ed. Seymour Chatman. *Literary Style: A Symposium.* New York: Oxford UP, 1971. 241-54.

Ong, Walter, J. *Ramus, Method and the Decay of Dialogue.* Cambridge: Oxford UP, 1958.

—. "The Writer's Audience is Always a Fiction." *PMLA* 90 (1975): 9-21.

"Outline of Principles and Purposes." *The Rhetoric Society Newsletter* 1 (1968): i-ii.

Perelman, Chaïm, and L. Olbrechts-Tyteca. *The New Rhetoric: A Treatise on Argumentation.* Trans. John Wilkinson and Purcell Weaver. Notre Dame, IN: Notre Dame P, 1969.

Perl, Sondra. "The Composing Processes of Unskilled College Writers." *Research in the Teaching of English* 13 (1979): 317-36.

Pike, Kenneth. *Language in Relation to a Unified Theory of the Structure of Human Behavior.* The Hague: Mouton, 1967.

Polanyi, Michael. *Personal Knowledge: Towards a Post-Critical Philosophy.* New York: Harper, 1958.

Pringle, Ian, and Aviva Freedman, eds. *Reinventing the Rhetorical Tradition.* Conway, AR: L & S, 1980.

Rhetoric Society Newsletter 1 (1968): 1-2.

Rhetoric Society Newsletter 1 (1969): 5-9.

Rohman, Gordon. "Pre-Writing: The Stage of Discovery in the Writing Process." *College Composition and Communication* 16 (1965): 106-12.

Rohman, Gordon, and Albert O. Wlecke. *Prewriting: The Construction and Application of Models for Concept Formation in Writing.* Cooperative Research Project #2174, Office of Education, US Department of Health, Education, and Welfare, 1964.

Scott, Robert L. "On Viewing Rhetoric as Epistemic." *Central States Speech Journal* 18 (1967): 9-16.

Shaughnessy, Mina. "Diving In: An Introduction to Basic Writing." *College Composition and Communication* 27 (1976): 234-39.

—. *Errors and Expectations.* New York: Oxford UP, 1978. 275-94.

Sommers, Nancy. "Revision in the Composing Process: A Case Study of College Freshman and Experienced Adult Writers." Diss. Boston U, 1978.

Steinmann, Martin, ed. *New Rhetorics.* New York: Scribner's, 1967.

"Students' Right to their Own Language." *College Composition and Communication* 25 (1974): 1-32.

Tade, George, Gary Tate, and Jim W. Corder. "For Sale, Lease, or Rent: A Curriculum for an Undergraduate Program in Rhetoric." *College Composition and Communication* 26 (1975): 20-23.

Tate, Gary, ed. *Teaching Composition: Ten Bibliographic Essays.* Fort Worth: Texas Christian UP, 1976.

Toulmin, Stephen. *The Uses of Argument.* Cambridge: Cambridge UP, 1969.

"The University of Tulsa Program in Rhetoric and Writing." *Rhetoric Society Newsletter* 3 (1973): 4-5.

Vygotsky, Lev. *Thought and Language.* Trans. Eugenia Haufmann and Gertrude Vakor. Cambridge, MA: MIT P, 1962.

Weathers, Winston. "Grammars of Style: New Options in Composition." *Freshman English News* 3 (1976): 1-18.

Weathers, Winston, and Otis Winchester. *Copy and Compose: A Guide to Prose Style.* Englewood Cliffs, NJ: Prentice, 1969.

Williams, Joseph. "Defining Complexity." *College English* 40 (1979): 595-609.

Winterowd, W. Ross. *Contemporary Rhetoric: A Conceptual Background with Readings.* New York: Harcourt, 1975.

—. "Dispositio: The Concept of Form in Discourse." *College Composition and Communication* 22 (1971): 39-45.

—. "The Grammar of Coherence." *College English* 31 (1979): 828-35.

—. *Rhetoric and Writing.* Boston: Allyn & Bacon, 1965.

—. *Rhetoric: A Synthesis.* New York: Holt, 1968.

Young, Richard, Alton Becker, and Kenneth Pike. *Rhetoric: Discovery and Change.* New York: Harcourt, 1970.

Young, Richard, and Alton Becker. "Toward a Modern Theory of Rhetoric." *Harvard Educational Review* 35 (1965): 450-68.

II

Theory-Building and Critiquing Corderian Rhetoric

ക ൃ

2

Toward a Corderian Theory of Rhetoric

James S. Baumlin
Southwest Missouri State University

Let us now sing the praises of famous men, our ancestors in their generations. . . . they were wise in their words of instruction; those who composed musical tunes, or put verses in writing. . . . Some of them have left behind a name, so that others declare their praise. But of others there is no memory . . . they have become as though they had never been born. . . .
—Ecclesiasticus (44:1, 5)

The saving grace of communication, wherein one gives by receiving or receives by giving, is the discovery of one's fellow man, of one's neighbor—that other self, in friendship or love, more real than myself because he is identified with the value discovered throughout meeting. Each gives the other essential hospitality in his better self; each recognizes the other and receives from him that same recognition without which human existence is impossible.
—Georges Gusdorf (Speaking)

There is language . . . because there is "the other."
—George Steiner (Real Presences)

And yet, and yet: I have written as much to hide as to reveal, have written so that I might show the writing to others and not be required to show myself. There's more to me than meets the eye, and less. Whatever is in here might be terrible to see, worse to reveal. A piece of writing can be revelatory, exploratory; more are than we have acknowledged. It can also be a substitute for the unspeakable, a closure, not a revelation.
—Jim W. Corder (Yonder)

☞In "Tribes and Displaced Persons" (1993), one of his late essays, Jim
W. Corder announced an informal retreat from the genres of "academic
writing." Paradoxically, Corder's rationale strikes at the heart of his early
rhetorical theory, which emphasizes self-revelation and the individual's
ethical relation to his or her writing and audience:

> I haven't yet learned how to be myself. . . . I don't want to learn
> how to be someone else. I can't be Maynard Mack, whose work
> I admired so much when I was in graduate school, and I
> probably won't turn out to be Jacques Derrida. I want to try to
> think my thoughts, which aren't altogether mine. I don't want
> to write in the languages of the academic communities I have
> almost belonged to for years. I hope this is the last piece that
> looks even a little like academic writing. . . . I want to do a
> scholarly sort of work but to write in a personal sort of way. . . .
> I want to write in my way, which isn't mine, and perhaps even
> stretch out the possibilities of prose. (281)

Pursuing "a scholarly sort of work . . . in a personal sort of way," Corder's
mature writing did, in fact, "stretch out the possibilities of prose," and
had he managed to complete, polish, and publish the remaining
manuscripts that he rushed to complete before his passing, his fame as
an essayist would endure beyond doubt. Yet Corder did not give up
entirely on "academic writing." Among his unpublished works is a
manuscript, "Rhetorics, Remnants, and Regrets," which seeks to
synthesize and expand upon his previous discussions of rhetorical
theory.[1] While passages are typically Corderian—that is, brilliant and
moving—and worthy of study, the work remains imperfect, again
reflecting the author's haste in finishing an important part of his life's
work. But, based on this manuscript, we can say that Corder was still
working toward a systematic theory of rhetoric, one that had been
developing over several decades and that had at least reached near-
completion.

What, then, shall be Corder's legacy in rhetorical theory? Much like
Robert M. Pirsig, whose *Zen and the Art of Motorcycle Maintenance: An
Inquiry into Values* (1974) interwove personal narrative with musings on
rhetoric, Corder successfully united autobiography and regional history
with serious explorations of rhetorical theory. Works such as *Yonder:
Life on the Far Side of Change* (1992)—arguably Corder's most powerful
collection—do present a uniquely Corderian rhetoric. But just as few now
read *Zen and the Art* as a serious exposition of theory, I fear that future
students of rhetoric will fail to look for the theory presented in Corder's
last published books. In short, Corder the theorist remains largely
submerged within Corder the belletrist—with serious implications for his
scholarly legacy. So, what would Corderian theory look like, had he
managed to complete and articulate it "in a scholarly sort of way"?

It should not surprise that Corder's thinking evolved. His earlier
writings—say, from 1970 to 1985—express an unwavering confidence in

the powers of language to reveal the speaking or writing self, to reduce conflict and create community, to accommodate diversity of belief and embrace competing truths; one might describe this as the "heroic" phase of Corder's thought, best represented by such texts as "Varieties of Ethical Argument" (1978), "From Rhetoric to Grace" (1984), and "Argument as Emergence" (1985). His later writings—roughly from 1986 to 1997—occasionally sound the same strains of confidence and linguistic optimism. For the most part, though, the works of this later, "tragic" phase view rhetoric as a crisis of ethos, communication, and accommodation.[2] Though the tragic elements of Corderian rhetoric grew out of the earlier, heroic phase and are perhaps the latter's inevitable destination, these two distinctive versions of Corderian theory are in most respects mirror opposites, a pair of mutually enabling (and yet, at the same time, competing) perspectives. It is not that Corder reversed his mind regarding rhetoric; rather, he came to appreciate the ways that any theory summons from within itself the potential for its own inversion, critique, and negation.

For, arguably throughout the history of rhetoric, each systematic theory has grounded itself upon distinctive premises regarding reality and the nature (or possibility) of truth; at the same time, each theory has rested upon premises regarding human nature and psychology, premises regarding culture and its effects upon individuals, and premises regarding the capacities (or limitations) and the various functionings of language. Is the truth knowable? Are authors present in or absent from their texts? Does a speaker own (or embody) his or her words, or are they the prior possession of culture? Is self-identity singular, stable, and univocal, or rather multivalent and multivocal? Is persuasion ethical? Is it even possible? The simple, unelaborated answers to such persistent questions include "yes and always," "no and never," "maybe sometimes," "perhaps rarely if ever," "it depends," "we cannot know." In his mature writings, Corder learned to vary his answers to the above questions, in each case invoking a different version or shape of his theory. Indeed, in his more stylistically experimental works, Corder often sought to explore a particular version and its *in*version simultaneously—as suggested by the essay title, "When (Do I/Shall I/May I/Must I/Is It Appropriate for Me to) (Say No To/Deny/Resist/Repudiate/Attack/Alter) Any (Poem/ Poet/Other/Piece of the World) for My Sake?" (1988).[3] Each rhetoric, thus, is describable as an epistemology, a psychology, a sociology of discourse, and a theory (indeed, an implicit theology) of language. And any change in any one of these underlying premises effects a corresponding change in the structures, powers, and functioning of discourse. This same recognition underlies Corder's notion of multiple, competing rhetorics—one of his greatest (if still controversial) contributions to theory.[4]

Through the following pages, I sketch out a broad (and, admittedly, partial) outline of Corderian rhetoric, seeking to reconstruct its major epistemological, psychological, and linguistic premises. I argue that Corderian expressionism is grounded squarely and demonstrably within an existentialist psychology, in which the problem of self-identity remains always the central focus. And it is for this reason that ethos, or "character as it emerges in language" ("Varieties" 2), dominates his theory: for Corder sought to redescribe the structures and processes of rhetoric as constitutive of human consciousness. I also observe that Corderian theory remains situated firmly within the canons of classical rhetoric (specifically invention, arrangement, style, and memory), thereby confirming the continuing utility of these terms to contemporary theory. As he writes in "From Rhetoric Into Other Studies" (1993),

> The vocabulary of rhetoric—*invention, structure, style, occasion, audience,* and other terms—is ubiquitous and inevitable and can be used in the discussion of diverse rhetorics: All statements come from somewhere, however knowable or undiscoverable, emerge as structures and styles, however deliberate or accidental, and occur on some occasion for some audience, however untimely, however small. Any line of inquiry, any field of interest, any subject matter, then, can be taken as a rhetoric or as a set of rhetorics. That, I think, makes it possible to use the vocabulary of rhetoric to discuss any human interest. (95)

Like his colleagues Edward P. J. Corbett, James L. Kinneavy, and Richard Weaver, Corder remained a devoted student of classical rhetoric. At the same time—and certainly more than his colleagues aforementioned—Corder sought to pour new wine into the old vessels, radically reinterpreting the classical canons and vocabulary. Even as we seek to understand, appreciate, and apply his theory, we can admit without much loss that Corder is intuitive, speculative, repetitive, often (by his own admission) quirky in his arguments, stingy in his scholarly citations, and given to formal and stylistic experimentation—all of which makes his writings, the later works especially, a mixed bag of mixed genres and effects. Clearly, Corderian theory requires careful scrutiny. Some of his more speculative ideas will prove difficult to test or apply; some of the more controversial may not stand. Yet the time is right for this revisiting, given that social-constructionism—still the dominant movement in contemporary composition theory—has more or less completed itself and stands itself in need of correction.

Shifting focus away from individual psychology and onto culture as the constitutive force of discourse, constructionist theory has offered a useful counterpoint to expressivist pedagogy. But, as Alan W. France and others have begun to argue, social-constructionism has erred in the opposite direction by denying the role (and, in the more extreme versions of theory, even the reality) of personal agency. For expressivists, France writes,

> have largely failed "to articulate the theories underlying their practices in any systematic way . . . while social constructionists were working out epistemological positions and promoting

> theoretical self consciousness" [O'Donnell 425]. But the inverse
> is true as well. For their part, social constructionists have not
> worked out much of a theory of personal agency, failing largely
> to offer a way out of the "disabling postmodern box of the
> always already determined subject" [Flannery 707]. (148-49)[5]

The claim that expressivists have "largely failed 'to articulate the theories underlying their practices'" needs qualification, though it is true that recent expressivism has lost sight of its own intellectual origins. These origins are not exclusively or even primarily "Romantic," as is often assumed; rather, as Kinneavy convincingly demonstrates in *A Theory of Discourse* (1971), they are to be found in existential phenomenology.[6] This should not surprise, given that existentialism defined postwar intellectual culture and remained a dominant philosophical movement from the late 1940s through the early 1970s (only then to be eclipsed by social-constructionism). Inspired by Jean-Paul Sartre's *What Is Literature?* (1947), "Why Write?" (1949), and other Sartrean reflections upon writing, an existential phenomenology of language was developed in such seminal works as Georges Gusdorf's *Speaking* (1965) and Maurice Merleau-Ponty's posthumous *The Prose of the World* (1973);[7] even our current use of the terms *expressivism* and *expressionism* seems to originate in these authors' emphasis upon language-as-expression.[8] As Gusdorf argues, personal expression is always "present as a coefficient of speech," so that "the whole of human experience in its militant sense may be understood as a striving for expression" (70, 73). Though the great urgency of humankind is to establish interpersonal relationship, nonetheless one's "relation to others is only meaningful insofar as it reveals that personal reality within the person who is himself speaking. To communicate, man *ex-presses* himself, that is, he actualizes himself, he creates from his own substance . . ." (69). Thus the function of expression, Gusdorf concludes, "consists in a movement of man outside himself in order to give meaning to the real. Expression is the act of man establishing himself in the world, in other words adding himself to the world" (71).

I am arguing, in fine, that Corderian expressionism remains an intellectual offspring of existential phenomenology—specifically, an extension of Gusdorfian language theory to writing instruction. As such, the theory underlying Corderian rhetoric is coherent, fully defensible, and an adequate counterpoint to current social-constructionism. Corder cites Gusdorf in "Varieties of Ethical Argument" and "From Rhetoric to Grace," and I would go so far as to suggest that the heroic phase of Corderian rhetoric is quintessentially a West Texas elaboration upon French phenomenology (and, concomitantly, that Corder's tragic phase is a questioning and critique of the same). This Sartrean, Gusdorfian influence by no means detracts from Corder's originality as a rhetorician; rather, it grounds his unique contributions—that include his notion of multiple rhetorics, his adequation of rhetoric and psychology and, perhaps most important, his theory of "generative ethos"—upon solid intellectual bedrock. It so happens that recent reactions against poststructuralism have reawakened interest in existentialism generally, especially as it offers a "humanist" alternative to the former's apparent antihumanism. And, humanist to the core, Corderian theory can help in this reawakening. As France observes, we have yet to synthesize social-constructionist, cognitivist, and expressivist pedagogies, though our profession does seem, salutarily, to be moving in this direction: Witness this very collection, which seeks to reassess the contributions of expressivism generally and Corderian rhetoric in particular. Following such reassessment, it is my belief that Corder will be recognized as one of the great American rhetoricians of the late twentieth century, the one who most fully realized the existentialist basis of expressionism and whose theory, thus, has much to teach us still.

CORDERIAN EXPRESSIONISM AS A MODE OF EXISTENTIALISM

Hazel E. Barnes describes the problematic self-identity of the Sartrean *pour-soi* or being for-itself: "as a lack of being, the for-itself reaches out toward being. Consciousness is not a self and does not have a self; but as a self-making process, it pursues a self. Or, as Sartre says, it seeks to come to itself" (143). This gives us a hint as to the faithful ethical relation aimed at in existentialist rhetoric, whether Sartrean, Gusdorfian, or Corderian: the preservation of the freedom of both speaker and audience and the refusal to coerce or manipulate. But there is a corollary, the basis of existentialist ethos being, as Sartre declares, a freedom "*to choose oneself*" (*Being* 538), a point to which Gusdorf returns: "[T]he presence of freedom . . . gives the human activity of speech its true dimension" (120). At its most optimistic, existentialism affirms the capacity of individuals to make (and remake) the self and emerge toward the other. Two points are thus of especial note: the grounding of discourse—that is, of "authentic" discourse—in the mutual freedom of both speaker and audience, and the role language plays in the processes of self-making. This emphasis upon authenticity derives from the existentialist conviction that "self-deception" is, as Abraham Kaplan puts it, "the greatest vice" (110):

> Life . . . is a drive towards honesty, toward really being what we
> are. Mere things . . . are wholly subject to the law of identity;
> for man, identity is something to be achieved. A man who is
> identical with himself—with his human self, the self which he
> has made by his free and responsible choices—such a man
> existentialism calls "authentic." . . . The man who lacks
> authenticity is indeed playing a role: his existence has yielded
> to an essence which defines what he is. (110-11)

Thus Kaplan invokes the central Sartrean notion that *existence precedes essence*. As an expression of "absolute freedom," the *pour-soi* can never be fixed within a singular role or stable, unified identity: more than an act of bad faith or self-deception, such would be a repudiation of the very freedom constitutive of human selfhood.[9]

In addition to freedom, our humanness is constituted by language. As Gusdorf writes, "if it be true that language is a world, it is also true that the world is a language that must conform to the influence of personal authenticity," for which reason "style is the peculiar expression of personality" (76).[10] Sartre's assertions are more radical still:

> In the intersubjectivity of the for-others, it is not necessary to invent language because it is already given in the recognition of the other. I *am* language. By the sole fact that whatever I may do, my acts freely conceived and executed, my projects launched towards my possibilities have outside of them a meaning which escapes me and which I experience. (*Being* 455)

Thus the French existentialist concludes that "Heidegger is right in declaring that *I am what I say*" (*Being* 455-56), to which Corder gives his assent: "Our words never leave us; the message is not separate from the speaker" ("Varieties" 20). Indeed, Corder is explicitly existentialist in asserting that the author "must be free in order to learn and to write,"[11] for authorship is a "perpetual hunt for texts only to back off, improvise, try again, search again through freedom to speak again in a continuous and ever provisional making of self and of world," by which means "we begin to learn how to cherish, to enfold the other, holding that other's need to make, to be free to speak, to be free to learn a new or deeper voice, to become a self always becoming a self" ("Jackleg Carpentry" 20).

Sartre is adamant in asserting the reader's freedom as well. Reading, indeed, becomes "a pact of generosity between author and reader" in which each is an equal participant in the making of meaning:

> Each one trusts the other; each one counts on the other, demands of the other as much as he demands of himself. For this confidence is itself generosity. Nothing can force the author to believe that his reader will use his freedom; nothing can force the reader to believe that the author has used his. Both of them make a free decision. . . . Thus, my freedom, by revealing itself, reveals the freedom of the other. . . . The writer appeals to the reader's freedom to collaborate in the production of his work. ("Why Write?" 1064)[12]

Given the antihumanist implications of poststructuralist thought, such heroic affirmations might sound hopelessly naive to some ears—as, for example, the following passage, where Corder describes his own ethics of reading and careful, caring response:

> If I criticize a single action or utterance of the other, that is the
> other's identity I speak against, not a separable item that may
> be regarded coolly, but an item that connects with, reverberates
> through, all of the other, hence *is* the other. If the other
> criticizes a single action or utterance of mine, that is my
> identity the other speaks against, not a separable item that may
> be regarded coolly, but an item that connects with, reverberates
> though, all of me, hence *is* me. ("When" 53)

Yet it should by now be clear that such an argument affirms the
Heideggerian/Sartrean claim that *"I am what I say."* American
theologian Arthur A. Vogel elaborates on this point:

> It is not that man is a being who uses words merely as
> something outside of him and external to him. The statement
> that man is linguistic means that his most immediate and
> intimate existence has the nature of a verbal utterance. . . .
> There is a sense in which man is a word; he does not just use
> words. In fact, man is able to use words only because his being
> has the nature of a word. (92)

By grounding language in intersubjectivity, existentialism thus
demands the equal, ethical copresence of speaker and audience, self and
other. As Corder writes in "Argument as Emergence," "we have *to see*
each other, *to know* each other, *to be present to* each other, *to embrace*
each other" (23). Such urgencies underlie Corder's best work, both early
and late. Time and again Corder would ask, not how does one persuade
but, rather, how does one establish contact? How can one come to be
heard? To be seen? How can one enter into another's inventive world?
How can one open his or her world to another, and at what risk? In S.
Michael Halloran's seminal essay, "On the End of Rhetoric, Classical and
Modern" (1975), he describes the sense of "fragmentation and isolation"
permeating postmodernism, a moment in history when ethos and ethical
appeal can succeed only to the degree that a speaker "is willing and able
to make his world open to the other," thus risking "self and world by a
rigorous and open articulation of them in the presence of the other" (627-
28). Elaborating upon this explicitly Sartrean argument,[13] Corder writes
that even as we are "apart from each other," and even as "distances open
between us," still "we keep trying to enter their world or bring them to
ours. Often we fail, but we keep trying":

> The trouble is that our speaking-forth—the primary need and
> issue of any age—is complex, confused, and messy, and often
> creates as many problems as it solves. Language is our way of
> composing ourselves. It is our first and last line of defense, and
> we are vulnerable in each line. ("Varieties" 2)

Following Halloran, Corder acknowledges the vulnerability and mutual
risk facing speakers and audiences, authors and readers alike, as well as

the need for developing a commodious discourse wherein language becomes the means, not just of "composing ourselves," but of making our "world open to the other"—opening a space, as it were, for the ethical copresence of self and other, as "we keep trying to enter their world or bring them into ours" (2).

Here, in "Varieties of Ethical Argument"—again an early essay, reflecting Corder's heroic theory—such risk-taking is vital, noble, and possibly successful: "often we fail," Corder admits, "but we keep trying." In his later, tragic phase, the sense of urgency in risk-taking is heightened; but so is the likelihood, not just of failure, but of a speaker's hesitation and even refusal to take risks—and, in Corder's personal narratives (his unpublished manuscripts included), such refusal is always accompanied by self-accusation. It would be an overstatement to suggest that Corder's later theory asserts the ineluctable, inevitable failure of intersubjectivity and ethical relation. Yet a work such as *Yonder* does recount a theorist's failure to abide by his own theory of communication. Though I have called it tragic, Corder's late work is actually quite courageous in its commitment to personal truth—even to the truth about lying, self-deception, guilt, and regret. I cannot, therefore, entirely lament the fact that Corder clothed his later theory in personal narrative. For he made, thereby, the ultimate commitment to expressionism. In effect, Corder the storyteller transforms Corderian theory into a story about Corder the theorist who tries to live by his own theory and tests it and laments his personal failure and dies before he can tell the story to a satisfactory completion, in the end proving that "the message is not separable from the speaker."

Thus, in accordance with existentialist expressionism, Corder viewed self-revelation and ethical relation as the great twin urgencies of communication, toward which one must strive and strive even as one fails and fails. And, in questioning the basis of a speaker's success or failure, we arrive at the intellectual core of Corderian theory: the role of ethos in establishing a commodious discourse. For it is by means of ethos that we establish our initial point of contact with the other, inviting the other into our inventive worlds. Hence the success—or failure—of communication is largely a consequence of ethos. And here, as I have suggested, Corder explores two contrasting versions of his theory.

THE VARIETIES OF CORDERIAN ETHOS

While acknowledging the classical Aristotelian definitions, Corder's own theory of ethos proceeds from the existential phenomenology that we have treated thus far. Within the classical discussions of ethical appeal, there is typically no sense of vulnerability or risk on the speaker's part, nor is the audience conceived as an equal, equally active participant. Yet the sort of ethos Corder imagines is jointly produced—a collaboration between speaker and audience—within a shareable inventive world,

which language opens up and sustains as a sacred space.[14] As Corder
writes in "From Rhetoric to Grace," "*the prime ethos is God*" (26): God
"becomes through a speaking forth in history, an ethos emerging through
the Word that we can imitate but not replicate. Given such a model,
however, we can both speak ourselves into existence and be with another
in language" (26). Thus God serves as the exemplar and guarantor—the
"divine anchorage" (xxix), as Merleau-Ponty's translator puts it—of
faithful ethical communication, the image of a divine "ethos emerging
through the Word." While the earlier "Varieties of Ethical Argument" (his
seminal and fullest discussion) is not explicitly theological, it too
describes an originary mythic or superhuman ethos whose presence is at
once sacred and assuring, though ultimately inimitable. Indeed, the
western landscape—particularly as depicted in Jack Schaefer's western
novel, *Shane*—provides the powerfully symbolic backdrop for this aspect
of Corderian existentialism: two sovereign presences crossing the empty
prairie in an archetypal confrontation between self and other.

As Corder's favorite literary illustration of heroic or "efficient ethos,"
Shane "fully uses his world, without waste. All that is in his inventive
realm realizes itself in an array of structures and styles so varied that
some of them reach over into use in another world" (14):

> Invention in Shane's world equals structure and style. The
> same thing, I think, might be said of Beowulf and of Robin
> Hood and of many of our heroes. The perfect economy of
> invention, structure, and style makes them heroic realities to
> us, and makes Shane a presence to the boy [narrator]. But this
> kind of ethos is self-completing—hence Shane's departure at
> the end. It is not self-renewing. Shane will be a memory, not a
> model for behavior. Neither Shane nor Robin Hood can live in
> a world devised by others. ("Varieties" 14)

Unlike the Sartrean *pour-soi*, the fully heroic or "efficient" ethos cannot
be remade; its greatness, fullness, completeness becomes, paradoxically,
a mark of its own fixity, its inability to be other than what it is. Corder
proceeds to describe several alternative versions, most important of
which is "generative ethos," a more approachable, humane form "that we
need both to hear in others and to make ourselves" ("Varieties" 14):

> Unlike the "efficient ethos" of Shane, which, heroic as it is, is
> tied to a particular inventive-structural-stylistic set and
> completes itself, the "generative ethos" is always in the process
> of making itself and of liberating hearers to make themselves.
> In this form of ethos there is always more coming. It is never
> over, never wholly fenced into the past. It is a speaking out
> from history into history. (14)

It is for this reason that "generative ethos" is "never completely achieved"
("From Rhetoric" 21). Like the Sartrean *pour-soi*, the Corderian
"generative ethos" realizes (or, rather, pursues) its humanness by

acknowledging its *lack* of being or fixed, stable essence, leading individuals to engage in a continual process of self-making and world-making. It is for this same reason that "argument is emergence toward the other, . . . a willingness to dramatize one's narrative in progress before the other" ("Argument as Emergence" 26):

> It is a risky revelation of the self, for the arguer is asking for an acknowledgment of his or her own identity, is asking a witness from the other. In argument, the arguer must plunge on alone, with no assurance of welcome from the other. . . . In argument, the arguer must, with no assurance, go out, inviting the other to enter a world that the arguer tries to make commodious, inviting the other to emerge as well, but with no assurance of kind or even thoughtful response. (26)[15]

As we have seen, Halloran and others had already affirmed this existentialist risk. Yet the great strength of Corderian theory lies in its extension of such insights into writing pedagogy. For generative ethos accords with the structures and processes of (self-)composition; that is, ethos describes not simply one's self-image or character but, rather, a living mode of rhetorical invention, arrangement, and style. And, as in the following passage, Corderian invention and style prove to be dialectically related:

> By its nature, invention asks us to open ourselves to the richness of creation, to plumb its depths, search its expanses, and track its chronologies. But the moment we speak or write, we are no longer open; we have chosen, whether deliberately or not, and so have closed ourselves off from some possibilities. Invention wants openness; structure and style demand closure. We are asked to be perpetually open and always closing. In this sense, rhetoric is a model of the examined life. If we stay open, we cannot speak or act; if we stay closed, we have succumbed to dogma and rigidity. ("From Rhetoric" 21)

Thus related, the processes of invention and style describe a modus vivendi as much as a *modus dicendi* or *scribendi*, both processes alternating as a sort of systole-diastole, a rhetorical heartbeat necessarily opening and closing, opening and closing, such alternation describing not just an aspect of writing but a condition of living human consciousness.

In "Varieties of Ethical Argument," Corder refers to "the self-authenticating language" of "generative ethos," whose aim is to invite audiences "into a commodious universe" (20). In so striving, the speaker must seek to delay or "shove back," as Corder puts it, the inevitable "closure" of language:

> When we speak, we stand somewhere, and our standing place makes both known and silent claims upon us. We make truth,

> if at all, out of what is incomplete, or partial. . . . Yet language is
> a closure, for we cannot speak two words simultaneously.
> Generative language seeks to shove back the restraints of
> closure, to make in language a commodious universe, to stretch
> words out beyond our private universes. (20-21)

This desire to delay closure, to "stretch words out beyond our private
universes," underlies the next great insight of Corderian theory, which is
that rhetoric—and for Corder, this encompasses all interpersonal
communication—is quintessentially an enactment and an expression of
love. In "From Rhetoric to Grace" (26), Corder quotes Gusdorf on the
relation between communication and "the communion of love." As
Gusdorf writes:

> True communication is the realization of a unity, . . . the unity
> of each with the other, but at the same time the unifying of
> each with himself, the rearrangement of personal life in the
> encounter with others. I cannot communicate as long as I do
> not try to bring to the other the profound sense of my being.
> The communion of love, which represents one of the most
> complete modes of understanding between two persons, can't
> be achieved without a recall of personality, each discovering
> himself in contact with the other. (57)

Though contemporary social-constructionists might find such sentiments
embarrassing, the existentialist assumption is that love, caring, or
"tenderness," as the German theologian Martin Buber puts it, underlies
all ethical relation.[16] For love, as Buber writes, "is responsibility of an I
for a You: in this consists what cannot consist in any feeling—the equality
of all lovers" (66). Premising his ethics of "tenderness" upon childhood
nurturing, Buber asserts "the *a priori* of relation" (79):

> In the drive for contact (originally, a drive for tactile contact,
> then also for optical contact with another being) the innate You
> comes to the fore quite soon, and it becomes ever clearer that
> the drive aims at reciprocity, at "tenderness." But it also
> determines the inventive drive which emerges later (the drive
> to produce things synthetically or, where that is not possible,
> analytically—through taking or tearing apart), and thus the
> product is "personified" and a "conversation" begins. The
> development of the child's soul is connected indissolubly with
> his craving for the You. (79)

Thus Buber articulates "a theory of meaning based on literal envisioning,
this is to say of the vision we have of the face, of the expressive 'thereness'
of the other human person" (Steiner 146), the rhetorical equivalent of
which would be the expressive "voice," whether intoned orally or
reconstituted in writing.[17]

In both its heroic and its tragic moments, Corderian ethos expresses
the same "craving for the You," the same urge toward self-making

mutuality, a soul-making communion offering all participants a chance for conversion, for "conversation," for health and growth and change. Yet "it is clear," too, as Noddings writes, "that my vulnerability is potentially increased when I care, for I can be hurt through the other as well as through myself" (33). It is, indeed, this very vulnerability that Corder explores in his later, tragic theory. We must keep in mind that classical rhetoric traditionally aims at persuasion or change: a change of mind and heart, of mood and motive, of intention and attitude, of belief and action. So, if rhetoric fails to effect change (whether in oneself or in others), it would seem to have failed in its essence. We must keep in mind, too, that the risk of opening to the other, described above, entails more than one's failure to change another: It marks the equal possibility that one might be changed in turn. The risk, again, is presumed to be mutual for both speaker and audience.

Radically existentialist, such issues remain powerfully relevant to Corder's later theory. As Noddings describes the psychology of guilt,

> Here am I, one who cared, who does not now care, and the other sees it. I can summon reason to my defense. Look at this other! What has he done to encourage or appreciate me? What a mess he is. How I have tried. . . . I can go on and on and guilt comes right along like my shadow.

> Can I avoid this? Can I be free of guilt? I do not think it is possible. Paul Tillich describes the anxiety of guilt as ontological. It transcends the subjective and the objective. It is a constant threat in caring. In caring, I am turned both outward (toward the other) and inward . . . when caring fails, I feel its loss. I want to care, but I do not. (38)

Following the German theologian Paul Tillich, Noddings treats the "anxiety of guilt" as ontological and, thus, as an ineluctable fact of human existence: Excepting perhaps saints and martyrs, we must each at times bear the anxious, burdensome guilt of uncaring. And not only is the risk of guilt "present in all caring . . . its likelihood is greater in caring that is sustained over time" (39).

Given the resolute openness of his earlier, heroic theory, it is not surprising that Corder's later writing confronts the inevitable corollary— the guilty recognition that, as one's caring changes, one's inventive world "closes" to the other. As Corder writes in *Yonder*:

> After the divorce: what she knows is not what I know; what I was is not what I am; what I am is not what I was; what I was is not what I was; what I am is not what I am; what I am is not here, floats somewhere—maybe here, maybe there. I have no witness and can't testify to my own existence. Won't. Don't know how. Might. Better not.
> Can't. (30)

Here as elsewhere throughout his later writing, personal narrative overshadows the underlying theory. But let us consider the following, less personalized paraphrase:

> What the other knows is not what I know;
> what I was is not what I am;
> what I am is not what I was;
> what I was is not what I was;
> what I am is not what I am.

In presenting this series of negations, Corder denies the capacity for self-making, self-revelation, and accommodation central to his earlier, heroic ethos. We might also remember a passage, previously cited, where Corder admits to having written "as much to hide as to reveal," for "whatever is in here might be terrible to see, worse to reveal. A piece of writing can be revelatory. . . . It can also be a substitute for the unspeakable." Thus he reverses the terms of his earlier theory, treating ethos as "a closure, not a revelation" (54).

In "Hiding," a chapter from *Yonder*, Corder recounts an event close to the time of his divorce and remarriage. Going to buy a Sunday newspaper, he caught sight of his daughter, Cathy: "She had, I think, just come out of the bookstore—she likes the Sunday *Times* too—and was getting into her car. I stepped quickly behind the corner of the building. Perhaps she saw me. Perhaps she didn't. I hid, at any rate. I hid" (208). As he continues, "I hid because I was ashamed, . . . hid because I thought you wouldn't want to see me" (210). Such an admission reverberates beyond the personal narrative, pointing to the crisis within Corder's theory of commodious discourse and of ethos as ethical self-revelation. Later in this chapter, he repudiates his former, idealized self-image for all of its inauthenticity (as Sartre would say, its "bad faith"), acknowledging at the same time that his previous "inventive world" had amounted to a willful imposition upon others, denying them freedom. "Oh, Cathy," he laments, "I thought all along that I was gentle and undemanding. In other eyes I was insistent, domineering, always wanting to define the world in my way, to claim it for myself, to hold it. I didn't compose well, and I misread my own text" (211).

And, too human, we are capable of depriving not just our fellows but ourselves of freedom. In "Jackleg Carpentry," Corder asserts our need to be "free to learn and to write, . . . to speak again in a continuous and ever provisional making of self and world" (20); otherwise, our lives and actions become inauthentic, a mere parody of the self-in-process. In this same heroic spirit, Gusdorf writes that we must each "contribute to creating a better world, a world prepared, announced, and even brought into being by every word that is a harbinger of good faith and authenticity" (127). But what if we lose—or, rather, give up—this freedom, this good faith, this commitment to personal authenticity? In "To the Jayton Cemetery," the penultimate chapter of *Yonder*, Corder describes his Baptist upbringing as a mental and emotional prison: "Why did you deprive me of Heaven, Preacher, you and all those others? Why did you and all those others—and I, oh yes and only—make a Heaven in my mind I cannot reach, a Hell I cannot miss?" (216-17). As a critical reflection upon Corder's earlier, heroic theory, this imagined loss of Heaven is no more (nor less) than a loss or disabling of one's powers of *invention*, of one's freedom and openness to revision and change. The loss, Corder admits, is entirely "in [the] mind," the effect of a mere story preached over and over by "an alien authority," as Corder would likely have described it in "Jackleg Carpentry," an authority whose demands for "conformity and imitation" must be resisted by all means ("Jackleg" 20). Painful and damning, yet it is a story become bred into his bones—a text-of-the-self that has become fixed, closed off from revision. And this, from an existentialist perspective, is the definitive human tragedy: the individual's loss of freedom and failure to remain a self-in-process.[18] In "Varieties of Ethical Argument," Corder quotes from Schaefer's novel: "[A] man is what he is," Shane tells us, "and there's no breaking the mold," to which Corder adds, "being what he is, Shane won't go into that new world" (13).

Shane cannot change. In the late essay, "Lessons Learned, Lessons Lost" (1995), Corder drives this point home: "[W]hen you learn strong lessons early, however wrong, no evidence seems to count against them" (15).

> How do you remember guilt, disgrace, honorable victory, honorable defeat, and success if the way you first learned them was maybe altogether wrong and certainly altogether mismatched to a world that any soul ever lived in? When persuasive people and daily evidence both testify otherwise, how do you continue to believe—and how wrong you would be if you did—that suffering is noble, that love is always accompanied by chivalric behavior, that the WASP family of 1934 is the appropriate goal of nostalgic dreaming, that true believers will at last be saved? (23)

While Corder did not live to articulate his full, final answer to these questions, he did leave a trail of suggestions throughout his later works. The truly Corderian answer, I believe, is that one must learn to reach beyond the fact of guilt and suffering in order to recover from one's failures and false assumptions and self-deceptions—self-recovery being a significant concept embedded in Corder's late theory. As he writes in *Yonder*,

> The man who believes he was born in sin and continuously damned thereafter will not use language in the same way that a woman will who believes she was born innocent or good with every possibility of better yet to come. The first can at best only use words to try *to recover* from his own limitations. The second, however, may be free to use words *to reveal* her own present character. (164)

The woman "born innocent" will express an ethos of self-realization and self-revelation, while the man "born in sin and continuously damned" is left to practice an ethos of self-recovery. The criterion generating (and separating) these distinctive versions of selfhood is only apparently theological, since Corder is not making any absolute truth-claims for either the innocence or sinfulness of the human creature; rather, he is demonstrating how one's language and style—one's rhetoric-as-lived—proceed from beliefs that shape not only one's world but, perhaps more important, one's self-image, which is itself subject to change over time.

In a sense, the Corderian heroic ethos is an expression (and, when successful, a fulfillment) of the openness and innocence of youth; though Corder does not explicitly make this connection, the heroic ethos is certainly youthful in spirit, self-making being the great task of adolescence and early adulthood. Youthful, we strive to make a self, to make a strong personality, to marry, to make a career—to grow into the hero of our own story. The Corderian tragic ethos, in contrast, awakens to the guilty recognitions and regrets of older age, when one has lived long enough to amass a personal history of failures and (self-)deceptions. And what happens upon recognizing that the story we have told of ourselves "was maybe altogether wrong and certainly altogether mismatched to a world that any soul ever lived in?" In that crisis of midlife and beyond, when our earlier, idealized self-image has proved to be little more than an occasionally pleasing fiction, are we not challenged to reconstruct a self-image truer to our new sadder-wiser understanding of the world? What will this reconstruction or "self-recovery" entail? For Corder, it necessitated not just the repudiation of his earlier, idealized self-image but the unmasking, debunking, and retelling of his entire life and times: hence the great personal narratives, *Lost in West Texas* (1988), *Chronicle of a Small Town* (1989), *Yonder* (1992), and *Hunting Lieutenant Chadbourne* (1993), along with the several unpublished manuscripts to which I have alluded.

I shall return to the subject of self-recovery, as well as to the theological ground of Corderian theory; here, though, one final aspect of Corderian ethos deserves attention. Even as he grew increasingly to acknowledge the ways that culture and ideology shape an individual's discourse, Corder remained unsettled (though unconvinced) by the poststructuralist effacement of authorial voice and presence. In this regard, two passages from *Yonder* are worth quoting at length, and both must be read in the light of Corder's commitment to the thorough unmasking, debunking, and retelling of history, both public and private:

> I had wanted to believe, you see, that *ethos*, the character of a speaker or writer, is real and dwells in the text left by the speaker; that it was there—just as Aristotle and, until yesterday, all who came after him had said. . . . Others, however, have been looking with different eyes and have consequently seen things differently. . . . Diverse others—those who have taught us recently about interpretive communities,

> about the social construction of discourse, about reader-response theory, for example—have told me in diverse ways that ethos is *not in the text* but in the reader or community, in their projection on the text. They have told me, further, that the writer is not autonomous, but is only part of a social community that constructs and interprets discourse, that the notion of the self as a source of meaning is only a Romantic concept . . . that the language by which we view and construct the world comes from society, not from the individual. *Ethos* . . . is at the very least problematic. I can't catch character in the text, it seems clear, not on my own previous terms. That may mean that I can't ever be real to you, or you to me, or the other to either of us. (53)

Thus the mature Corder acknowledges the poststructuralist refutation of his own existentialist ethos of self-revelation, the theory to which he had devoted decades of a distinguished academic career. The passage immediately following is a rather ambivalent elaboration upon this same refutation, half accepting and half resistant:

> We have been, whether knowingly or not, whether directly or not, part of a twenty-five-hundred-year-old tradition that allowed and encouraged us to believe that character is in the text, that authors do exist, that they can be in their words and own them even in the act of giving them away. Now literary theorists both compelling and influential tell us that it is not so, that character exists, if at all, only in the perceiving minds of readers; that authors, if they exist, do so somewhere else, not in their words, which have already been interpreted by their new owners. Language is orphaned from its speaker: what we once thought was happening has been disrupted. Authors, first distanced, now fade away into nothing. Not even ghosts, they are projections cast by readers. "They" out there want, that is to say, to take my own voice away from me and give such meaning as there might be over entirely to whoever might show up as an interpreter. "They" want me—and you and all those others—to die into oblivion if we should manage to scribble something, never to be reborn in a voice for some reader, but to vanish before that reader's construction/deconstruction/reconstruction of the small things we leave behind. (54)

So we "die into oblivion, . . . never to be reborn in a voice": Bowing to the death-knell pronounced by Roland Barthes, Michel Foucault, and Jacques Derrida among other poststructuralists, Corder thus laments the apparent loss of authorial presence (and of incarnationist rhetoric generally).[19]

And yet, consistent to the end, Corder sought not simply to theorize but, rather, to live through the poststructuralist deconstruction of voice and self-identity, giving his own authorial death and disappearance an intensely personal expression:

> We can never find exactly who we were or whatever might have
> been. Whoever we were and whatever might have been have
> already been lost to us in the transformations of compressing
> and expanding rhetorics.
>
> There—nostalgic among competing rhetorics with failing
> memory—I'm always on the point of disappearing. As it is
> always possible for me to be compressed, perhaps to nothing,
> as it is always possible for nostalgia to lose me elsewhere, as I
> cannot depend on memory to recall who I am, invisibility is
> always imminent, especially in the presence of all those others
> who seem to know, who seem to know both text and territory.
> (*Yonder* 172)

Here in *Yonder*, Corder's most expansive meditation upon postmodern
ethos, the author's voice and very being stand "on the point of vanishing."
For, according to "theorists both compelling and influential," there was
never a stable, abiding "autonomous self" to reveal after all. In "Notes on
a Rhetoric of Regret," Corder expresses his personal sadness and
"lonesomeness" over the apparent loss:

> We have generally been slow to know ourselves as multi-vocal
> among multiple rhetorics. I am a little sad, remembering that
> old idea, that old hope for an integrated, autonomous self.
> Whatever it is that I want to imagine I am or might have been
> is already lost, unheard, unwitnessed, unnoticed, and I am
> lonesome for the self that was never present. (95)

Throughout his later writings, Corder turned this same "lonesomeness"
into an active search for a transcendent self, other, and world, a search
(or "hunting," as he was wont to term it) that would lead him down the
literary-generic pathways of regional history, biography, and personal
narrative.[20]

For, however diminutive or inconsequential, each individual life
possesses its own grandeur; in proof of which, Corder chose to recover
and retell the brief, heroic-tragic life history of Theodore Lincoln
Chadbourne, a second Lieutenant "killed on May 9, 1846, in the Battle of
Resaca de la Palma at the opening of the Mexican War" ("Hunting
Lieutenant Chadbourne" 343). The young man's story seems at first
glance unremarkable, "Link" Chadbourne being but twenty-three "when
he died that day" (345). But it is precisely Chadbourne's perilous
marginality, his near-disappearance from history, that attracted the
mature Corder's passionate attention: Like Corder himself, who feared
(and predicted, even as he resisted) his own inevitable disappearance,
Chadbourne stood in need of a living witness. As Corder confesses, "I
wanted to find Lieutenant Chadbourne" ("Hunting Lieutenant
Chadbourne" 347).[21] And Corder wished to believe, did believe, that he
had found Chadbourne, if only partially, that he had found him not so
much in the few precious museum artifacts—the young lieutenant's

sword, sash, and battle-torn waist coat—and not so much in the scant historical records as in the small collection of private letters that Corder himself had personally, doggedly, hunted down. While summarizing their contents, Corder paused to offer the following admonition, delivered as an elaborate parenthesis:

> (Listen, literary theorist—you who say the author is not the author but only one in a social community that actually writes the text, you who say that the character of the speaker is not in the text, but is only my projection, you who say that the author dies when the reader is born—come be with me in my life for a while, follow me as I transcribe his letters, watch me wince at the first letters, then shrink at the later letters, then shake at the typewriter as I read his last letters. . . . Of course I never wholly found him. But of course I did, too, and I didn't want him to die. But he died, and I cannot be rid of his presence.) (358)

Come be with me in my life for a while: Such is the plea of Corderian rhetoric, both early and late.[22] But poststructuralist theory questions whether such intersubjectivity is even possible. By what means, then, can Corder legitimately claim to have found Lieutenant Chadbourne, much less himself?

GOD, THE GROUND OF CORDERIAN THEORY

Grounded upon a divine ontology, Corder's rhetorical theory proceeds from (and expands upon) the Judeo-Christian existentialism of Martin Buber, Teillard de Chardin, Paul Tillich, and Elie Wiesel, among other theologians, for whom the presence or absence of God—humanity's great and enabling, if problematic, "Other"—remains the ethical-intellectual crisis of our postwar, postholocaust, postmodern age. As with other aspects of his theory, Corder's divine ontology flies in the teeth of current intellectual fashion, given that we live, as Steiner notes, "in a moment of history where the frankly theological" is "largely held in derision" (223); and yet the very "category of meaningfulness" (225) remains theological in essence, for "the Hebraic intuition that God is capable of all speech-acts except that of monologue . . . has generated our arts of reply, of questioning and counter-creation" (225). By preserving God as the ground and guarantor of truth, meaning, and being, Corder sought an antidote to deconstruction's negative or "zero theology" of language.[23] In so doing, Corderian rhetoric—like other historical versions of theory—proves as much a theology as a psychology and a sociology of language. There should be no need to apologize for this aspect of Corderian theory: Since its classical-Hellenic inception, Western rhetoric has remained in dialogue with the varieties of Western theology, just as it has interacted with the varieties of Western philosophy, ethics, politics, science, psychology, and aesthetics. Indeed, to be complete and fully articulated, any systematic theory must identify the full range of its enabling premises, which include its attitude toward ultimate, transcendent being. Deconstruction, thus, proceeds from a linguistic atheism in which God's absence leads not just to the loss of meaning (in effect divorcing the word from the world) but even to the loss of self, other, and world as stable, knowable, transcendent categories of being.

As George Steiner asserts in *Real Presences* (1989), "any coherent account of the capacity of human speech to communicate meaning and feeling is, in the final analysis, underwritten by the assumption of God's presence" (3). Indeed, human history itself is enabled by a similar faith *in language*, in the faithful wedding of the word to the world, such that "any . . . deconstruction of the individuation of the human speaker or *persona* is, in the context of Western consciousness, a denial of the theological possibility and of the *Logos* concept which is pivotal to that possibility" (99).[24] By this re-wedding, Steiner's existentialism seeks also to preserve reading and writing from deconstructive claims of authorial absence. Just as speech affirms the bodily presence of its participants, so reading describes a similarly existentialist confrontation, if not with the being then with the "offered *meaning*" of the author/other (156). By affirming this abiding presence-of-meaning, Steiner describes a mode of authorial presence resistant to deconstruction, at the same time preserving the author's ethical relation to his or her texts. But it must be noted that, like all linguistic theologies, Steiner's system rests upon an initial act of faith: Just as all truth-claims proceed from a prior belief in the possibility of truth, so Steiner's theory of communication rests upon a prior assumption of meaning-fullness in language (119). Being the prior assumption of a systematic theory, such a claim can be acknowledged, discussed, and defended but never proved from within (or by means of) the theory itself. And any theory remains most vulnerable at its foundation; in this sense, deconstruction is nothing more than an inversion of the epistemological, linguistic—and, yes, theological— premises upon which expressionist theory is based, replacing the latter's existentialist affirmations with its own radical skepticism.

I emphasize this effect of systematic theory because it demonstrates the ways that expressionism remains vitally engaged with poststructuralist thought: Put simply, the deconstructionist's skepticism requires positive faith as its enabling "other," as even the very statement, "I doubt," makes sense only in a world where "I believe" remains an equal possibility. By recognizing and exploiting this fact, Steiner saves not only his own expressionist theory but Corder's as well; at the very least, he shows us how to mount an adequate defense of Corderianism. For Steiner's "gamble on welcome" is very much in the spirit of Corder's "emergent ethos":

> Face to face with the presence of offered meaning which we call
> a text . . . we seek to hear its language. As we would that of the
> elect stranger coming towards us. There is in this endeavor, as

> deconstruction would immediately point out, an ultimately
> unprovable hope and presumption of sense, a presumption
> that intelligibility is conceivable and, indeed, realizable. Such a
> presumption is always susceptible of refutation. . . . In short,
> the movement towards reception and apprehension does
> embody an initial, fundamental act of trust. It entails the risk of
> disappointment or worse. But without the gamble on welcome,
> no door can be opened when freedom knocks. (*Real Presences*
> 156)

The similarities between Steiner's and Corder's thought are patent,
though I do not argue for any direct influence; my point, rather, is that
Corder's late lamentations over the loss of selfhood must not be taken as
the final, definitive version of Corderian theory. For Corder's tragic
phase, as I have been suggesting, is no more than a deconstruction of his
earlier, heroic theory; and Steiner teaches us how vitally the
deconstructionist's skepticism requires its enabling other. Here, too, the
Corderian "I doubt," makes sense only where "I believe" remains an
equal possibility.

Let me confess that my aim throughout has been to save the
Corderian legacy, to reassert Corderian heroism even in the midst of
tragedy, to reintroduce Corder to Corder and to keep both self-images,
both versions of Corderian ethos, in living conversation. The Corderian
hero was born innocent, expressing an ethos of self-realization and self-
revelation; the Corderian scapegoat was born in sin and continuously
damned. Let the former help heal the latter, leading him to self-recovery,
helping him end his restless hunting for self, other, and world. And let
the divine ethos guarantee that finding of self, other, and world, thereby
transforming Corderian tragedy into its rightful, higher, divine comedy.
The heroic phase of Corderian rhetoric finds perhaps its fullest, most
forceful expression in the essay, "From Rhetoric to Grace" (1984), from
which I shall quote two passages; by now, I doubt that they need much
commentary:

> We cannot be God speaking all into being throughout history,
> but it is possible—if the stars are right and we work to make
> ourselves human—to enfold another whose history we have not
> shared. In this act of enfolding, the speaker becomes through
> speech; the speaker's identity is always to be saved, to emerge
> as an *ethos* to the other, whose identity must also be cherished.
> Then we may speak to another, each holding the other wholly
> in mind. (26)

> Beyond any speaker's bound inventive world lies another: there
> lie the riches of creation, the ineffable presence of Knowing—
> indeed, God—all the great unbounded universe of invention.
> All time is there, past, present, and future. The natural and the
> supernatural are there. All creation—God and the knowledge of
> God—is there, ground and source for invention. The knowledge
> we have is formed out of the plenitude of creation, which is all

before us, but must be sought again and again through the cycling process of rhetoric, closing to speak, opening again to invent again. In an unlimited universe, pregnant with meaning to be made, we can never foreclose on interpretation and speaking. Invention is a name for a great miracle—the attempt to unbind time, to loosen the capacities of the past, the present, and the future into our speaking. (26)

As Corder concludes this essay, "invention is God, the site and origin of grace" (27): Our human structures and styles "are closures, but we have the plenitude to return to, grace that offers delight and salvation in new knowledge.... We are never fully in grace; we are reaching toward grace. It is not earned; neither is it given. It is *there*, waiting.... The language we use is never enough, but we get to return, if we will, and say more" (27). Thus Corder equates communication with communion, a sacrament of language grounded against a divine backdrop of infinite inventive possibilities, wherein the limitations of the human creature—symbolized, once again, by "closure," by our inability to speak more than one word, one world, at a time—are compensated for by God's eternal openness and by the gracious gift of second and third chances, whereby we can unspeak and respeak again and again until we get things right. It is by means of such blessing that Corder's tragic ethos yields to a higher, divine comedy of forgiveness and recovered freedom.

Having cited passages from this earlier, heroic essay, it is only fair to let the later Corder speak in turn. "Losing Out" (1994) remains, I suspect, a largely unknown and unread piece, in which Corder both acknowledges and seeks to move beyond the poststructuralist assault against selfhood. Here, truly, the tragic Corderian reaches across the chasm of deconstruction to shake hands with the earlier hero. But further commentary must cease, existentialist, deconstructionist, or otherwise: Corder himself must have the final word. For it is to his strong and, ultimately, hopeful voice (as well as to the other strong voices that I have cited: Gusdorf, Merleau-Ponty, and Steiner especially) that we may once again turn after we have finished with the latest passing fashion, needing to relearn what Corder never ceased teaching us. Speaking ourselves into being before the gracious presence of God and our fellows, we remain responsible for the worlds that we make:

> I am not a single, unitary, stable personality. I have composed myself and revised myself, always a little nostalgic for that other self I could not compose, the one that sounds like Adrienne Rich, maybe, or Jacques Derrida, but looks like Gregory Peck. I am provisional and plural. Any writer or speaker is always plural, though the responsible *I* is always singular. (99)

> And yet . . . we are still here, responsible for ourselves though doubting ourselves. We keep trying to tell our souls to the world. We leave tracks. (99)

Yet think of the grand gifts we won. Think of what is perhaps the grandest of all: we became individual souls, lost and alone in the world to make the world, becoming self-conscious, no matter how quirky, how mired in ideology; we created perspective; we made the world. We can be mean, vengeful, and narrow, but we must remember to say, too, that we are precious and dear and momentous. We are what we have; we are what the world has. We are the world's agents of inquiry, discovery, and creation, lost in error, capable of magic. There are no others, just ourselves. (100)

NOTES

[1]Listing the following "work in progress," Corder's last vita (compiled 1991) includes "Competing Rhetorics and Human Behavior," "first draft complete, revisions under way." By late 1997 this work had grown to 342 double-spaced type-script pages and assumed the title, "Rhetorics, Remnants, and Regrets: An Essay on Shifting and Competing Rhetorics." The final draft shows considerable rearrangement of pages and paragraphs, including numerous handwritten interpolations. Corder did in fact send the manuscript to an academic press, receiving a letter of rejection some weeks before his passing. (I have not seen the press readers' reports but have been told that the rejection proved a blow to his spirits.) Along with "Places in the Mind" and "Scrapbook," two other unpublished manuscripts, this manuscript remains in the private possession of his surviving family.

[2]The very titles of Corder's later essays often suggest their antiheroic spirit: "The Time the Calvary Didn't Come" (1985), "Lonesomeness in English Studies" (1985), "The Heroes Have Gone from the Grocery Store" (1987), "Lamentations For—and Hopes Against—Authority in Education" (1988), "Hunting for *Ethos* Where They Say It Can't be Found" (1989), "Academic Jargon and Soul-Searching Drivel" (1991), "Lessons Learned, Lessons Lost" (1992), "At Last Report, I Was Still Here" (1993), "Losing Out" (1994), "Notes on a Rhetoric of Regret" (1995).

[3]And how, we might pause to ask, are we to read this title? Note how the phrases within each set of parentheses establish the various transposable forms of a complex question. The title asks to be read as an algorithm, set up as a spatial grid allowing for each transmutation of the question to be present at once. In one version the question reads, "When Must I Resist a Poet . . . ?" In another, "When Is It Appropriate for Me to Alter Any Piece of the World . . . ?" By such experimental grammar, Corder sought a means to overcome the severe temporal limitations of language—the fact that thought fills our mind in all its immediacy and simultaneity, whereas the words by which we offer our thought to another must be spoken (or, more slowly, written) linearly and sequentially, one at a time. For "each sentence," as Georges Gusdorf tells

us, "orients us in a world which . . . is not given as such, once and for all, but appears to be constructed word by word" (37). In contrast, "total expression would be the actualization of all possibilities, the liberation of all the competitors for being that make up a personal reality—a kind of unravelling of man" (87) and, for this reason, humanly impossible.

Unlike the open-endedness and generativity of thought, one's speaking-forth is, as Corder notes, necessarily "a closure, for we cannot speak two words simultaneously" ("Varieties" 21), a crucial point to which Corder returns time and again:

> We are always standing somewhere when we speak. Whenever we use language, some choices have already been made and other choices must be made. Our past accompanies us into our statements and exercises its influence whether or not we are entirely aware of it. Before we speak, we have lived; when we speak, we must continually choose because our mouths will not say two words simultaneously and our hand will not write two words simultaneously. Whether consciously or not, we always station ourselves somewhere when we use language. That means that what rhetoricians call *invention* always occurs. ("From Rhetoric" 17)

Let me add that, as much as possible, I wish to let Corder's own texts speak for themselves; hence I shall quote his more important and influential works at length.

4As Corder writes in *Yonder*, "each of us is a gathering place for a host of rhetorical universes" (165):

> Some of them we inhabit alone, and some of them we occupy without knowing that we do so. Each of us is a busy corner where multiple rhetorical universes intersect. Each of us is a cosmos. All rhetorical universes are still present. The debris of all our generations is always around us and in us. Sometimes they cross or try to mix. Sometimes we don't remember them well or keep order among them. Each of us is a crowd. Into this gathering—diverse personal rhetorics, culture rhetorics, folk rhetorics, family rhetorics, emotional rhetorics, physiological rhetorics, and on, and on—uncertainty and change keep introducing disorder. (165-96)

5France (by the way, a student of Corder's) quotes from O'Donnell's "Politics and Ordinary Language: A Defense of Expressivist Rhetorics" (1996) and Kathryn Flannery's "Composing and the Question of Agency" (1991). Sherrie Gradin argues similarly: "[F]or the past decade we have been trapped in an adversarial relation between expressive or romantic and social-epistemic theories and their proponents" (403), during which time the latter theories have become increasingly self-promoting and rigidly doctrinaire, refusing to "look to other theories that might still [prove] valuable" (403). Please note, by the way, Gradin's equation of "expressive" with "romantic" theory, a point to which I shall soon return.

[6]Even its proponents accept unquestioningly the reputedly Romantic origins of American expressivism. In *Reason to Believe: Romanticism, Pragmatism, and the Possibility of Teaching* (1998), for example, Kate Ronald and Hephzibah Roskelly trace its origins directly to Emerson and Thoreau, while treating expressivism and Romanticism as largely equivalent, interchangeable terms. In contrast, Kinneavy observes that "concern for the expressive function of language is a modern phenomenon" (394); as he argues in "Expressive Discourse" (393-449), a chapter from his *Theory of Discourse*, the existential phenomenology of Sartre, Gusdorf, and Merleau-Ponty "represents a serious attempt to reintegrate culture, to give reality a meaning, and to integrate 'other' and the self into culture" (397). In so writing, Kinneavy echoes Gusdorf's claim that "for each of us language accompanies the creation of the world—it is the agent of that creation" (*Speaking* 39):

> It is by speaking that man comes into the world and the world comes into thought. Speaking manifests the being of the world, the being of man, and the being of thought. All spoken words, even in negative or self-deceptive speech, attests to the horizons of thought and the world. (39-40)

Though neglected by contemporary students—for a salutary exception, see Thomas Newkirk's *The Performance of Self in Student Writing* (1997)—Kinneavy's *A Theory of Discourse* remains the seminal discussion of the intellectual premises underlying expressivist pedagogy.

[7]The original French edition of Gusdorf's *La Parole* appeared in 1959, Merleau-Ponty's *La Prose du monde* in 1969 (though the latter, evidently a fragment of a larger projected text, had been completed by 1952). Both were translated as part of the distinguished scholarly series, Northwestern Studies in Phenomenology and Existential Philosophy. Taken together, they represent a coherent and fully adequate expressionist theory of communication. As Alla Bozarth-Campbell summarizes Gusdorfian expressionism:

> Gusdorf articulates a phenomenology of speaking as a form of bodily gesture expressing time (the convergence of past and future in the *now* moment of utterance) and space (the speaker's place in the world). Not speech, but the speaking human being in a context of human reality is what constitutes meaning and reveals experience. For Gusdorf speaking implies a nondualistic attitude toward the world: the person as lived experience, speaking as a way of being-in-the-world and stretching out towards others. . . . Speaking is always a process and a relation. (57)

These same themes, we shall see, reappear throughout Corder's writings. And, though probably not a direct influence upon Corder's work, Merleau-Ponty provides an additional foundation for expressionist theory generally, as Bozarth-Campbell suggests:

> For Merleau-Ponty language varies and amplifies intercorporeal communication, as it speaks thoughts from a body-subject to a body-subject, causes the other to speak and to become what he or she *is* but could never have been alone. Thus, we do not have speech, speech has us. We are formed by, become by, means of our speech. Speech as a gesture moves beyond thought, beyond self, to fill and create a *beyond*. . . . The body-subject knows itself only in encounter with the world and through other body-subjects. (67)

Note, then, the grounding of expressionist theory in lived experience and bodily presence. It is also, perhaps, not too soon to note that the poststructuralist effacement of authorial "presence" makes a direct assault against existential phenomenology and its expressionist theory of language. As John O'Neill, Merleau-Ponty's English translator, wrote, "the phenomenological approach to language is ultimately an introduction to the ontology of the world. It is a reflection upon our being-in-the-world through embodiment" (xxxi).

[8]Gusdorf's *Speaking* includes such chapter titles as "Speaking as the Threshold of the Human World," "Speaking and the Gods: Theology of Language," "Speaking as Encounter," "Expression," "The Authenticity of Communication," and "Toward an Ethic of Speech," while Merleau-Ponty's *Prose of the World* includes such chapters as "Science and the Experience of Expression," "Dialogue and the Perception of the Other," and "Expression and the Child's Drawing."

[9]Kaplan is patently more optimistic than Sartre regarding our human capacities for self-making: Much like the Jungian notion of individuation, toward which each individual strives (but proves rarely able to reach), the task of self-making remains for most of us a process rather than an achievement. And it must be so: As we shall see, for Sartre—as for Corder—any claim to personal authenticity demands that one remains ever a self-in-process.

[10]As Gusdorf continues, "each of us is charged with realizing himself in a language, a personal echo of the language of all which represents his contribution to the human world," for which reason "the struggle for style is the struggle for consciousness" (76). Corder argues similarly:

> Everyone has a style in this sense; everyone is a style in this sense; not everyone knows that he or she has or is a style. Sometimes we unwittingly lock ourselves into styles; sometimes we do so out of arrogance or ignorance or dogma. Style understood in this sense need not, however, persist unchanged, for style has another meaning. *Style,* or better, *styles,* is a name for enabling capacities—things we learn that let us do other things, methods we acquire, abilities we acquire. Style understood in this sense can free or enlarge style taken as personal identity. However, if a person adopts a style (as enabling capacity) without genuinely making it his or her own, without learning from it, then style in this sense becomes affectation. ("From Rhetoric" 19)

[11]Having had the honor of coauthoring "Jackleg Carpentry" with my dear friend and colleague, I can attest to Corder's ownership of this and the following quotations. But let me assert, at the same time, the existentialist provenance of "jack-leggery," whose numerous spiritual cousins include the nautical term *jerry-rigging* as well as the bricolage of French structural-anthropologist Claude Lévi-Strauss. As Corder writes, we want things "to stay fixed . . . we want, perhaps need, a post to hitch ourselves to so that we won't get loose" (18):

> But we're always loose. For that reason, . . . we want to celebrate the bricoleur's trade, known in West Texas as jackleg carpentry . . . that mode in which, upon completion of a job, the carpenter backs off, surveys the work, and says, "well, there it is, by God,—it ain't much, but it'll hold us until we can think of something better." (18)

Corder's intuition here is correct: As Merleau-Ponty's translator observes, "expression is always an act of self-improvisation in which we borrow from the world, from others, and from our own past efforts" (xxxiv).

[12]As Merleau-Ponty writes, "I am not active only when speaking; rather, I precede my thought in the listener. I am not passive while I am listening; rather, I speak according to . . . what the other is saying. Speaking is not just my own initiative, listening is not submitting to the initiative of the other, because as speaking subjects we are continuing, we are resuming a common effort more ancient than we, upon which we are grafted to one another and which is the manifestation, the growth, of truth" (143-44). Recently, George Steiner argued similarly: In the meaning-making of reading, one "invests his own being in the process of interpretation. His readings, his enactments of chosen meanings and values, are not those of external survey. They are a commitment at risk, a response which is, in the root sense, responsible" (8). Thus Steiner outlines an existentialist ethics of interpretation:

> Interpretative response under pressure of enactment I shall [call] *answerability*. The authentic experience of understanding, when we are spoken to by another human being or by a poem, is one of responding responsibility. We are answerable to the text . . . in a very specific sense, at once moral, spiritual and psychological. (8)

[13]In private correspondence Halloran confirms his essay's existentialist basis. Drawn from the first chapter of his dissertation, *A Rhetoric of the Absurd* (1973), Halloran's "debts [are] to Sartre, Kierkegaard, and Camus." As his students will attest, this is an essay Corder regularly cited and discussed.

[14]I am arguing that Corder, like Gusdorf, acknowledges the "sacred character" (*Speaking* 121) of language and the fact that speech "at its most effective takes on the meaning of a *vow* or even a *sacrament*. It is

speaking in action, a word that is a holy action, a moment of personal eschatology in which destiny is shaped" (123). I shall have more to say regarding the linguistic theology underlying Corderian rhetoric.

[15]Let me add that Gusdorf anticipates Corder's own fondness for the term *emergence*. As Gusdorf wrote, "to speak is to wake up, to move toward the world and others. Speaking actualizes an emergence" (*Speaking* 93-94).

[16]As Nodding describes the "ethics of caring":

> When I care, when I receive the other. . . . there is more than feeling; there is also a motivational shift. My motive energy flows toward the other and perhaps, although not necessarily, toward his ends. I do not relinquish myself; I cannot excuse myself for what I do. But I allow my motive energy to be shared; I put it at the service of the other. It is clear that my vulnerability is potentially increased when I care, for I can be hurt through the other as well as through myself. But my strength and my hope are also increased, for if I am weakened, this other, which is part of me, may remain strong and insistent. (33)

[17]We often describe written texts as having voice, and expressivist pedagogy has traditionally aimed at teaching students to find or fashion the same; indeed, one of Corder's early rhetoric anthologies is titled *Finding a Voice* (1973), and the term is often used interchangeably with ethos. And yet, etymologically, the Latin *vox* means no more than *verbum* or "word" *in its spoken form*: that is, *vox* is that sound which is a spoken word, though we tend now to identify it with the word's speaker and his or her manner of speech. From an existentialist perspective, voice describes an activity of human bodily consciousness, making use of the material organs of speech and marking the essential unity of a word with its speaker. To say that writing has "voice" is, thus, to resort to an incarnationist metaphor in which the text hypostatizes its author's living, breathing speech. But, arguably, all discourse is oriented toward (or proceeds from) the body; in this sense, ethos can be equated with the material, bodily presence "standing before" the texts that it speaks or writes. Articulating such a view, Vogel suggests that words are indeed "extensions of the body," a sort of "meaning in matter, a location of presence"—literally an embodied presence. For meaning is in words, Vogel argues, "as we are in our bodies, and it is only because we are our bodies that we can 'be' our words—or, as it is usually put, mean what we say. We can stand behind our words because our presence overflows them and is more than they can contain, but we choose to stand behind them with our infinite presence because we are also in them" (92).

[18]I return to "Jackleg Carpentry":

> [I]t is process and growth, and not perfection or conformity to some standard, that makes discourse true and beautiful and personally useful and personally valuable. . . . It seems that

> discourse must always be jacklegged. It must, for its own
> health, never be "fixed," never perfect or complete. . . . It must
> only (and always) be "good enough," capable of change, always
> in motion. . . . Other things we can own, even admire; only the
> jacklegged things can be continually present and loved and
> involved in our being. (20)

And yet "the jacklegged discourse that is us (and our students) will always die before an alien authority" (20); hence the fully authentic ethos is "generative" and "emergent," the ethos of the self-in-process.

[19]"Writing," as Barthes claims, "is that obliquity into which our subject flees, the black-and-white where all identity is lost, beginning with the very identity of the body that writes" (49); for, where "writing begins, the voice loses its origin, the author enters into his own death" (49). Often cited and discussed, the poststructuralist "death of the author" hardly needs elaboration here, though I would point to Barthes' explicit rejection of incarnationism on behalf of a disembodied hermeneutic of signs—his denial, in short, of "the very identity of the body that writes." It is here, over the problem of bodily presence and consciousness, that existentialism and poststructuralism wage their fiercest intellectual battle.

[20]The themes of loss and loneliness are expressed in such Corderian titles as "On the Way, Perhaps, to a New Rhetoric, But Not There Yet, and If We Do Get There, There Won't Be There Anymore" (1985), "Lonesomeness in English Studies" (1986), "The Heroes Have Gone from the Grocery Store" (1987), *Lost in West Texas* (1988), "I Can't Get Away from Hoppy" (1990), "Lessons Learned, Lessons Lost" (1992), *Yonder* (1992), "Losing Out" (1993), "The Time the Calvary Didn't Come" (1985), and "Notes on a Rhetoric of Regret" (1995). The alternative themes of searching or hunting, and—possibly—finding, are expressed in "Hoping for Essays" (1989), "Asking for a Text and Trying to Learn It" (1989), "Hunting for *Ethos* Where They Say It Can't Be Found" (1989), "At Last Report I was Still Here" (1993), *Hunting Lieutenant Chadbourne* (1993), and "Hunting Lieutenant Chadbourne: A Search for Ethos Whether Real or Pretended" (1994).

[21]As its full argument defies summary, I shall settle for an extensive quotation of a minor passage—no more than an aside or diversion, as Corder admits below—in hope that its poignant immediacy will attract readers to the whole of this remarkable essay:

> But I may not find Lieutenant Chadbourne. He was still
> twenty-three when he died that day. Twenty-three years
> doesn't give a fellow much time to compile records or to write
> diaries or to get into history books.
>   But I have been looking. I corrected the errors I made
> initially and set out on his trail, though I have been often
> diverted.
>   For example, as I was beginning to learn a little, I
> consulted the *Register of Graduates and Former Cadets of the*

United States Military Academy. Thirty-nine young men were
graduated in the class of 1843 with Theodore Lincoln
Chadbourne. One of them was U. S. Grant. Nineteen of them
eventually rose to the rank of general, if not in the United
States Army then in the Confederate Army. But what seized my
attention was that thirteen of them died before they were forty,
and they died in strange places—Contreras, Monterey, Fort
Leavenworth, Resaca de la Palma. I started looking at other
classes, all from 1842 to 1861, and found that the astonishing
death rate persisted. In those twenty years, well over eight
hundred young men were graduated from West Point. A third
of them died young. That knowledge did not, however,
minimize the death of any single one, or diminish its jolting
impact. I felt colder when I saw Lieutenant Chadbourne in that
company. All those young men. ("Hunting Lieutenant
Chadbourne" 345-46)

²²And what are we to say of Corder's additional claim, "I cannot be
rid of his presence"? The essay's final paragraphs acknowledge the
possibility of intersubjective presence, though they offer no further
guarantee:

In the letters that survive, I found no trace of deceit or rancor.
He is never self-serving, never mean-spirited. He is a gentle,
funny, honest, strong young man, with a wide-ranging capacity
for care.
    And he slips away from me. Why wouldn't he? I can't
remember who *I* was five years ago, let alone who somebody
else remembers from five years ago. He slips away from me,
and doesn't. What of the other letters, I wonder, those I didn't
see? What of himself, whom I didn't see, and did? ("Hunting
Lieutenant Chadbourne" 365)

²³As Steiner observes, deconstruction declares nothing less than the
loss or destruction of interpersonal communication and of "communion"
generally, since "the 'other'" is presumed to have "withdrawn from the
incarnate,"

leaving either uncertain spores or an emptiness which echoes
still with the vibrance of departure. . . . Within Derridean
readings lies a "zero theology" of the "always absent." The *Ur*-
text is "there," but made insignificant by a primordial act of
absence. (229)

²⁴Steiner's elaboration is worth quoting in full:

[D]econstruction knows that there is in each and every
assumption of a correspondence (however subject to skeptical
and epistemological query) between word and world, in each
and every previous rhetoric of direct or indirect
communication and reciprocal intelligibility between speakers,

between writers and readers, a declared or undeclared delusion, an innocence or political-aesthetic cunning. The ultimate basis of such delusion, innocence or cunning, its final validation, are theological. Where it is consequent, deconstruction rules that the very concept of *meaning-fullness*, of a congruence, even problematic, between the signifier and the signified, is theological. . . . The archetypal paradigm of all affirmations of sense and of the significant plenitude—the fullness of meaning in the word—is a *Logos*-model. (119)

WORKS CITED

Barnes, Hazel E. "Sartre's Concept of the Self." *Critical Essays on Jean-Paul Sartre*. Boston: G. K. Hall, 1988. 137-60.

Barthes, Roland. "The Death of the Author." *The Rustle of Language*. Trans. Richard Howard. New York: Hill and Wang, 1986. 49-55.

Baumlin, James S., and Jim W. Corder. "Jackleg Carpentry and the Fall from Freedom to Authority in Writing." *Freshman English News* 18 (1990): 18-25.

Bozarth-Campbell, Alla. *The Word's Body: An Incarnational Aesthetic of Interpretation*. Tuscaloosa: U of Alabama P, 1979.

Buber, Martin. *I and Thou*. Trans. Walter Kaufmann. New York: Scribner's, 1970.

Corder, Jim W. "Academic Jargon and Soul-Searching Drivel." *Rhetoric Review* 9 (1991): 314-26.

—. "Argument as Emergence, Rhetoric as Love." *Rhetoric Review* 4 (1985): 16-32.

—. "Asking for a Text and Trying to Learn It." *Encountering Student Texts*. Ed. Bruce Lawson, Susan Sterr Ryan, and W. Ross Winterowd. Urbana, IL: NCTE, 1989. 89-98.

—. "At Last Report I Was Still Here." *The Subject Is Writing*. Ed. Wendy Bishop. Portsmouth, NH: Boynton/Cook, 1993. 261-66.

—. *Chronicle of a Small Town*. College Station: Texas A&M UP, 1989.

—. "From Rhetoric into Other Studies." *Defining the New Rhetorics*. Ed. Theresa Enos and Stuart C. Brown. Newbury Park, CA: Sage, 1993. 95-108.

—. "From Rhetoric to Grace: Propositions 55-81 About Rhetoric, Propositions 1-54 and 82 et seq. Being as Yet Unstated; or, Getting from the Classroom to the World." *Rhetoric Society Quarterly* 14.3/4 (1984): 15-28.

—. "The Heroes Have Gone from the Grocery Store." *Arete: The Journal of Sports Literature* 5 (1987): 73-78.

—. "Hoping for Essays." *Literary Nonfiction: Theory, Criticism, Pedagogy*. Ed. Chris Anderson. Carbondale: Southern Illinois UP, 1989.

—. "Hunting for *Ethos* Where They Say It Can't Be Found." *Rhetoric Review* 7 (1989): 299-316.

—. *Hunting Lieutenant Chadbourne.* Athens: U of Georgia P, 1993.

—. "Hunting Lieutenant Chadbourne: A Search for Ethos Whether Real or Pretended." *Ethos: New Essays in Rhetorical and Critical Theory.* Ed. James S. Baumlin and Tita French Baumlin. Dallas: Southern Methodist UP, 1994. 343-65.

—. "I Can't Get Away from Hoppy." *New Mexico Humanities Review* 33 (1990): 107-13.

—. "Lessons Learned, Lessons Lost." *Georgia Review* 46 (1992): 15-26.

—. "Losing Out." *Diversity: A Journal of Multicultural Issues* 1 (1993): 97-100.

—. *Lost in West Texas.* College Station: Texas A&M UP, 1988.

—. "Notes on a Rhetoric of Regret." *Composition Studies: Freshman English News* 25 (1995): 94-105.

—. "On the Way, Perhaps, to a New Rhetoric, But Not There Yet, and If We Get There, There Won't Be There Anymore." *College English* 47 (1985): 162-70.

—. "Places in the Mind: Essays on Rhetorical Sites." Unpublished manuscript, 1997.

—. "Rhetorics, Remnants, and Regrets: An Essay on Shifting and Competing Rhetorics." Unpublished manuscript, 1997.

—. "Scrapbook." Unpublished manuscript, 1997.

—. "The Time the Calvary Didn't Come, or the Quest for a Saving Authority in Recent Studies in Higher Education." *Liberal Education* 71.4 (1985): 305-19.

—. "Tribes and Displaced Persons: Some Observations on Collaboration." *Theory and Practice in the Teaching of Writing: Rethinking the Discipline.* Ed. Lee Odell. Carbondale: Southern Illinois UP, 1993. 271-88.

—. "Varieties of Ethical Argument, With Some Account of the Significance of *Ethos* in the Teaching of Composition." *Freshman English News* 6 (1978): 1-23.

—. "When (Do I/Shall I/May I/Must I/Is It Appropriate for me to) (Say No To/Deny/ Resist/Repudiate/Attack/Alter) Any (Poem/ Poet/ Other/Piece of the World) for My Sake?" *Rhetoric Society Quarterly* 13 (1988): 49-68.

—. *Yonder: Life on the Far Side of Change.* Athens: U of Georgia P, 1992.

Corder, Jim W., ed. *Finding a Voice.* Glenview, IL: Scott, Foresman, 1973.

Corder, Jim W., and James S. Baumlin. "Lamentations for—and Hopes against—Authority in Education." *Educational Theory* 38 (1988): 11-26.

—. "Lonesomeness in English Studies." *ADE Bulletin* 85 (1986): 36-39.

Flannery, Kathryn. "Composing and the Question of Agency." *College English* 53 (1991): 701-13.

France, Alan W. "Dialectics of Self: Structure and Agency as the Subject of English." *College English* 63 (2000): 145-65.

Gradin, Sherrie. "Revitalizing Romantics, Pragmatics, and Possibilities for Teaching." *College English* 62 (2000): 403-07.
Gusdorf, Georges. *Speaking (La Parole)*. Trans. Paul T. Brockelman. Northwestern Studies in Phenomenology and Existential Philosophy. Evanston, IL: Northwestern UP, 1965.
Halloran, S. Michael. "A Rhetoric of the Absurd: The Use of Language in the Plays of Samuel Beckett." Diss., Rensselaer Polytechnic Institute, 1973.
—. "On the End of Rhetoric, Classical and Modern." *College English* 35 (1975): 621-31.
Kaplan, Abraham. *The New World of Philosophy*. New York: Vintage, 1961.
Kinneavy, James L. *A Theory of Discourse*. New York: Norton, 1971.
Merleau-Ponty, Maurice. *The Prose of the World*. Ed. Claude Lefort. Trans. John O'Neill. Northwestern University Studies in Phenomenology and Existential Philosophy. Evanston, IL: Northwestern UP, 1973.
Newkirk, Thomas. *The Performance of Self in Student Writing*. Portsmouth, NH: Boynton/Cook, 1997.
Noddings, Nel. *Caring: A Feminine Approach to Ethics and Moral Education*. Berkeley: U of California P, 1984.
O'Donnell, Thomas G. "Politics and Ordinary Language: A Defense of Expressivist Rhetorics." *College English* 58 (1996): 423-39.
Pirsig, Robert M. *Zen and the Art of Motorcycle Maintenance: An Inquiry into Values*. New York: Morrow, 1974.
Ronald, Kate, and Hephzibah Roskelly. *Reason to Believe: Romanticism, Pragmatism, and the Possibility of Teaching*. Albany: State U of New York P, 1998.
Sartre, Jean-Paul. *Being and Nothingness: An Essay on Phenomenological Ontology*. Trans. Hazel E. Barnes. New York: Washington Square, 1953.
—. *Qu'est-ce que la littérature? What is Literature?* Trans. Bernard Frechtman. London: Methuen, 1965.
—. "Why Write?" *Critical Theory Since Plato*. Ed. Hazard Adams. New York: Harcourt, 1970. 1058-68.
Schaefer, Jack. *Shane*. Boston: Houghton, 1949.
Steiner, George. *Real Presences*. Chicago: U of Chicago P, 1989.
Vogel, Arthur A. *Body Theology: God's Presence in Man's World*. New York: Harper, 1973.

3
Jim Corder's Radical, Feminist Rhetoric

Keith D. Miller
Arizona State University

ᔓBetween 1961, the year his first scholarly essay appeared in *College English*, and his death in 1998, Jim Corder published literally dozens of essays in national scholarly journals; numerous essays in national scholarly collections; and dozens of other essays and poems in regional and local journals, magazines, and books. He also authored, coauthored, and edited five well-received textbooks, including a handbook issued in six editions; a book aimed at college English teachers and university administrators; and five other volumes.[1] At least three additional Corder books await posthumous publication.[2]

Over four decades Corder wrote at least as much—if not more—than anyone else in the entire field of rhetoric and composition.[3] But a paradox is at work, for other scholars do not cite Corder very much. In fact, an estimable figure in composition studies can write a history of composition teaching during the period of Corder's career without mentioning him at all. Joseph Harris did just that.[4] Sometimes Corder definitely appears to be a "vacancy," a word he used to describe himself ("Notes" 104).

Of course, quantity of publication does not necessarily equal quality. But consider that Corder won the first Richard Braddock Award offered by *College Composition and Communication*. And consider that virtually all the best editors in Rhetoric and Composition—Edward Corbett, Gary Tate, Richard Larson, Lee Odell, Theresa Enos, George Yoos, Victor Vitanza, Wendy Bishop, and others—published his work in such journals as *College English, Publication of the Modern Language Association, Quarterly Journal of Speech, PRE/TEXT, Rhetoric Review*, and *Freshman English News/Composition Studies*, and in collections. Why did Corder both appeal to editors and then tend, as he wrote, to "disappear"?[5] I suggest that this paradox stems from his reliance on a form of academic argumentation that was highly unorthodox, especially before the 1990s.

While Corder doesn't call himself a radical, a pioneer, a subversive, or a feminist, his rhetoric is radical, pioneering, subversive, and feminist. He interrogates, overturns, and supplants the agonistic rhetoric of display that dominates scholarly writing (and much of Western culture), replacing it with a feminist rhetoric. (Of course, he generates one of many feminist rhetorics, not *the* feminist rhetoric, which does not exist.)

Through a variety of strategies, Corder enacts six large principles of argument.[6] These principles and strategies dominate his scholarly discourse, especially his academic publications between 1971 and 1993, and together form a definable system of argument that defamiliarizes and subverts standard academic discourse by creating a sense of puzzling over a problem with a reader instead of handing her solutions. By highlighting his own uncertainties, Corder invites a reader to wrestle with issues, aware that her conclusions may differ from his. Paradoxically, he also asserts four unqualified claims—rhetoric is an architectonic discipline, ethos is the most important element in persuasion, truth is contingent (or "jacklegged"), and good rhetoric usually requires gentleness.

Focusing chiefly on essays that Corder wrote for prominent national journals of rhetoric and composition, I attempt to explain the main principles and strategies of his rhetoric. I hope this account will help elucidate his contribution to the theory and practice of argumentation—an elucidation necessitated in part by his own refusal to theorize, explicate, clarify, or even acknowledge much of his own system of argument. This refusal itself constitutes part of his argument. Following this analysis I elaborate my contention that his rhetoric is radical, pioneering, subversive, and feminist.

Here are the six principles and their strategies:

I. ARGUMENT AS INDIRECTION

Instead of presenting scholarly argument as an overwhelming, watertight display of knowledge, Corder argues indirectly through the following strategies:

Circling

Although Corder never mentions modern Spanish philosophy, Jose Ortega y Gasset's statement of philosophical strategy also articulates one of Corder's rhetorical tactics:

> [W]e will move steadily ahead toward a goal which I will not now spell out because it would not yet be understood. We will go moving toward it in concentric circles. . . . The great philosophical problems demand a tactic like that which the Hebrews used for the taking of Jericho and its innermost rose gardens: making no direct attack, circling slowly around them. (17-18)

Near the end of *Uses of Rhetoric*, Corder announces that he "backed off" crucial issues (208), perhaps because his strategy meant "taking the long way around, expecting it to be the shortest way home" (118). Four pages into "I'll Trade," he declares, "I'll start again" (33).

In "A New Introduction to Psychoanalysis, Taken as a Version of Modern Rhetoric," Corder circles repeatedly. He begins by referring warmly to Abraham Lincoln, the lunar landing, and a bevy of canonical literary figures who seem quite unrelated except for their irrelevance to both psychoanalysis and rhetoric. Three pages into the essay, he circles, "I have not yet begun, but I want to stop and come at things in another way so that I can sneak up on what may be my subject from another direction" (140). Then he begins to observe that mental illnesses embody blocked or misdirected Invention, Structure, or Style (which I am tempted to call the Holy Trinity of Corder's rhetoric, but won't). Nineteen pages into the essay, he circles again, "I expect I've talked long enough about originating thoughts. I should get to what I came for" (156). Two pages later he circles once more, "Now I can go on with what I have wanted to propose from the start" (159). Then, having finally returned to the proposition announced in his title, he elaborates it.

Building Clotheslines

In "Learning the Text: Little Notes on Interpretation," Corder mentions a phone call to his mother in which she emerged a larger, more complete person than he realized. Then he examines the meaning of *oratio* and its parts, each of which, he declares, transcends its standard textbook definition. After developing and illustrating this point, he finally proposes an unlikely analogy: Just as his mother exceeded the "parent text" that he "created" for her, so do *narratio* and *exordium* "blossom" when they escape their "confinement in textbooks" (244, 248). Here the apparently divergent discussions (of his mother and *oratio*) meet at the end, when he offers his surprising argument by analogy.

"Learning the Text" follows what nonfiction writer John McPhee calls a "clothesline" structure (qtd. in Roundy 73-74). Starting from two far-away points—his mother and *oratio*—Corder develops a clothesline for each, hanging different comments on each line. At the end, when he presents his analogy, the two clotheslines meet to form a "V."

In another essay Corder highlights his family's hardships during the Great Depression, mentioning incidents of boys throwing rocks at occupied outhouses. Then he notes the public demand that English teachers go "back to the basics." Developing a clothesline for each topic, he pulls the lines together when he argues by analogy that returning to the "basics" of English teaching would resemble returning to the poverty of the 1930s ("Outhouses"). Clothesline arguments are indirect because the two topics initially appear unrelated.

Bouncing

In "Learning the Text," when Corder completes his comments on his mother, he bounces to the seemingly unrelated subject of *oratio*. When he ends the essay with his analogy, he completes the "V" of his clothesline and validates bouncing from his mother to *oratio*.

Corder builds another clothesline in "Studying Rhetoric and Teaching School." Early in the essay, he recounts mowing his yard and then bounces to the history of black-eyed peas, sweet potatoes, turnips, and radishes—all of which "once were weeds" (9). He next bounces to English textbooks, some of which he finds authoritarian. Then he compares the process of weeds-to-potatoes to "the cyclical, naturally-replenishing nature of Invention, Structure, and Style" (29). By proposing the unexpected analogy between weeds-to-potatoes and the composing process, he completes his clothesline. Uniting what seemed two radically disparate subjects makes his bounce from potatoes to rhetoric seem highly purposeful.

But whereas "Learning the Text" features one bounce (from his mother to *oratio*) and the "V" of a simple clothesline, "Studying Rhetoric" includes not only weeds-to-potatoes and composing but also memories of lawn-mowing, rage at dictatorial textbooks, and a reflection on Sears catalogs and their role in rural West Texas. Veering back to English studies, he bemoans "reductionist, monistic [teaching] practices" that manifest "arrogance or ignorance or dogma" (17). Only then does he bounce back to complete his clothesline by proposing his surprising analogy between weeds-to-potatoes and composing. Complexity arises because his lawnmower and the Sears catalog assuredly do not hang from either clothesline. Careening yet again, he offers a six-page account of ethos before alluding to Lincoln and making a final series of bounces to the Hebrew Bible; the Christian Bible; and works by Oliver Goldsmith, Alexander Pope, Samuel Johnson, James Kinneavy, and Walter Ong. None of these texts hangs from a clothesline either. Despite its clothesline, "Studying Rhetoric" features enough bouncing to defy any notion of conventional organization. So do many of Corder's other works. By abandoning prefabricated structures in favor of bouncing, Corder generates momentum by creating the impression of thinking aloud.

Reveling in Self-Contradiction

Aristotle founded his logic on the principle of consistency. Corder, however, disorients readers by contradicting himself. In "Hunting for *Ethos*," Corder dismisses an earlier essay, calling it "frequently uninformed" (299). In "Hunting Lieutenant Chadbourne" he apologizes for misrepresenting Chadbourne in his earlier *Lost in West Texas*.

Corder sometimes contradicts himself within a specific work. For example, in many ruminations about rural West Texas, he portrays a weather-beaten childhood during the painful 1930s when a profound yearning for a boy's innocence wars against hardscrabble memories. He presents an unresolvable tension between nostalgia and antinostalgia toward a childhood wondrous and woebegone. Deprivation constantly jostles wistfulness.[7]

The title of one essay telegraphs self-contradiction: "On the Way, Perhaps, to a New Rhetoric, but Not There Yet, and If We Do Get There, There Won't Be There Anymore." Other examples abound. In "A New Introduction to Psychoanalysis," he declares, "In my earlier efforts to think, which I have tried to record in previous pages, there was some presumption of normality. Now I believe that . . . there is no normality" (157). In "Hoping for Essays," he first mentions "personal essays (and other kinds, if there are other kinds)" (311). Next he presumes that there are other kinds: "I don't know what a personal essay is or how it differs from other forms of writing" (313). Soon he reverses himself: "Every piece of [nonfiction prose] writing is a personal essay" (313). In "Notes on a Rhetoric of Regret," "Asking," and "At Last Report," he explains the need to pour one's self into one's writing but yokes that need to the realization that every text is a construction, not the author's selfhood. "Hunting Lieutenant Chadbourne" and the book of that title assert that Chadbourne, a soldier who died in the Mexican War, is and isn't in his letters, just as Corder is and isn't in his texts about Chadbourne. The tensions never ease.

In a radical move, Corder installs the anti-Aristotelian topos of self-contradiction as the linchpin of another essay. On the first page, he asks, "When may I privilege my way of seeing and thinking? Answer: Never. Answer: Always" ("When" 49). Elaborating, he explains the necessity of criticizing others and the violence of doing so: "I *always* count myself, *always* render judgments, *always* sanctify my judgments, *always* privilege my own way of seeing and thinking." But, he quickly counters, "I must learn *never* to count myself, *never* to deny another, *never* to render a judgment against another, *never* to sanctify my judgments, *never* to privilege my own way of seeing and thinking" (50). This profound contradiction structures the entire essay.

Exploding Taxonomical Boxes

Corder once told me: "All taxonomies leak." But instead of constructing a better ship with tighter compartments, he places dynamite into the leaking holes of every available taxonomy and genre category. Exploding cargo-holds disorients readers by disrupting their expectations, which hinge on the stability of well-defined genres.

Corder hints at this deconstructive project in a 1967 essay, one of the extremely few formal, non-Corder-like scholarly pieces that he ever produced. In a model interpretation of a poem by John Dryden (which he explicates as a species of classical rhetoric), he lets slip the comment that an unspecified "we" are "aware of the marriage of poetic and rhetoric" ("Religion" 248). Of course, many failed to recognize and bless such a marriage, but he did.

The first chapter of *Uses of Rhetoric* weds rhetoric to poetic as Corder analyzes a magazine ad, a TV show, an essay by Paul Roberts, and Tennyson's "Ulysses." In subsequent chapters he extends the marriage by examining Aristotle on ethos, a poem by Robert Browning, a poem by Oliver Goldsmith, an essay by Samuel Johnson, and current composition textbooks—among many other topics. In "Varieties of Ethical Argument," he anatomizes *Gulliver's Travels*, Aristotle's account of ethos, Otis Walter, Michael Halloran, Cicero, Jack Schaefer's *Shane*, the Gettysburg Address, a poem by Alexander Pope, the biblical Book of Amos, and the biblical Book of First Corinthians—among other topics. One form of ethos or another, he maintains, imbues all the literary works (and the speech) that he discusses. Here the varieties of ethos are distinguishable, rhetoric and poetic are not. In "Rhetoric and Literary Study" he assails literary critics as "a few priests" who worship "holy tablets" and proposes that rhetoric serve as an umbrella for literary studies (13, 19).

Corder constantly torpedoes genre distinctions between personal essays and scholarship, occasionally bemoaning others' failure to blend the expressive with the intellectual ("Academic Jargon"). While such demolition might satisfy others, it serves as a mere preamble for Corder, the radical, who blasts walls that separate entire fields. Then he merges those disciplines. In "Hunting Lieutenant Chadbourne" and the book by the same title, he fuses autobiography, postmodern rhetoric, and nineteenth-century American historiography. Elsewhere, dismantling a seemingly insuperable barrier between rhetoric and brain science, he defines psychoanalysis as a rhetoric, outlines a rhetoric of cognition, and yearns for a "biorhetoric" to explain mental illness ("A New Introduction," "On the Way"). Transgressing another, well-policed disciplinary border, he explicates cancer as extremely destructive, cellular crossovers of Invention, Structure, and Style ("On Cancer"). To raid different genres and disciplines is to argue indirectly by prodding readers to ask: "What is going on?"

Refusing to Theorize or Explicate

Corder seldom admits that he is arguing indirectly. He never explains why. For example, in "A New Introduction to Psychoanalysis," he repeatedly reports that he is circling but never supplies a reason. In none of his essays, textbooks, or other books does he directly theorize or advocate Argument by Indirection and the virtues of circling, building clotheslines, bouncing, contradicting one's self, or exploding taxonomical boxes. Ironically, his textbooks recommend formal, conventional types of argument, types that he constantly defies. His refusal to theorize, propagandize, clarify, and elaborate his Argument by Indirection—this refusal itself helps constitute his Argument by Indirection.

All Corder's strategies of Argument by Indirection invite a reader to share his process of contemplating and weighing. Just as he patiently explores a problem, he implicitly argues, so should a reader. A cardinal advantage of Argument by Indirection is that unlike the codified textbook formalism that he rebelled against, his strategies enshrine the rich waywardness of words and thought while, paradoxically, propelling thought through disciplinary fences.

II. ARGUMENT AS MYSTERY

As if the strategies of his Argument as Indirection were insufficiently confounding, Corder adds Argument as Mystery, spicing his mix of personal and academic writing with enigma, elegy, oracle, incantation, vision, and prophecy.

Even though the Double Mountains are so tiny that they do not appear on most maps of Texas, Corder repeatedly proclaims: "God lives on the Double Mountains." He never explains this "theology."[8] Does God reside elsewhere, too? Does God ever travel? If the Double Mountains loom magnificently in the internalized landscape of his childhood, do others internalize similar landscapes? Or is the statement merely whimsical? Because Corder never says, his comment remains enigmatic. Inasmuch as many of his writings about West Texas are, in part, elegiac, his declaration of "theology" may be elegiac as well; for he writes in Fort Worth, at a distance from the Double Mountains and his early childhood, when he lived near them.

Corder supplies arguments for his claim that the usual academic disciplines are rhetorics. But he also defines "painting"—an art of lines and colors, not words—as "a rhetoric" ("On the Way" 164). Because he fails to explicate this declaration, it never becomes a formal or an informal argument. It is oracular.

Corder ends *Uses of Rhetoric*, a decidedly academic book, by evoking a Whitman-like, mystical union with nature: "When I waded in a tributary of Duck Creek," he states, "I became Duck Creek." When "my bent-feather scout friend" departed the Brazos River, "he became, all those years ago, the shape of the rock I threw" (209). This passage is incantatory.

In 1971 Corder advocated not writing across the curriculum (WAC), but rhetoric across the curriculum—the entire curriculum (*Uses*). He contended that each course in each discipline should be organized according to Invention, Structure, or Style. This visionary proposal anticipated the writing-across-the-curriculum movement that emerged the following decade, but was far more revolutionary than WAC. It remains an unfulfilled vision.

In "For Sale, Rent or Lease," a 1975 essay in *College Composition and Communication*, George Tade, Gary Tate, and Corder describe the undergraduate rhetoric major that they proposed at their university only to be rebuffed. Their essay outlines their plan and offers it to others. A wonderful possibility that might be adopted somewhere, sometime, this vision generally has yet to arrive.

In 1985 Corder predicted an electronic revolution in which small machines would house thousands of widely available texts: "To make a new rhetoric we will have to face the implications of miniaturization and electronic communication and to decide whether new technologies may indeed bring a new kind of literacy and with it a new kind of rhetoric" ("On the Way" 165). High-reduction photography, he continues, allows one

> to put 3200 pages on a single ultramicrofiche film card. When that miniaturizing capacity is hooked to networks made possible by the computer, a significant library can be held in small space at reduced cost, and almost anyone almost anywhere can easily gain access to almost anything printed. ("On the Way" 166)

Of course, this prophecy proved accurate.

In the same essay, Corder calls for the analysis of visual images welded to texts. Rhetoricians, Corder insists, "will have to learn to understand the interplay of visual image and verbal message that makes a meaning of arguments as display in such places as television and magazine advertisements and on billboards" ("On the Way" 164). This prophetic statement anticipates the most recent thought of Gunther Kress and John Trimbur.

III. ARGUMENT AS CHARM

Rather than the common scholarly practice of besieging readers with claims and proof, Corder beguiles them by presenting a simple persona and by salting his writing with humor. Developing a folksy persona enables Corder to mitigate the disorientation that readers experience when encountering his Argument as Indirection. Even when examining or inventing a postmodern puzzle, he generally eschews jargon, explaining that he would rather plug away than operationalize ("You"). After *Uses of Rhetoric* appeared in 1971, he discarded the frequent practice of incorporating innumerable quotations, boxcarlike, into academic prose. Instead of supplying Argument as Boxcar, he cites judiciously, sometimes sparely. Further, his "ah-shucks" persona seems to reflect the rural boyhood that he frequently ponders in his scholarship and his later identity as a professor with one-third of a red bandana poking out of his hip pocket.

In "Some of What I Learned at a Rhetoric Conference," the simple persona reigns as Corder refuses the role of eager listener who absorbs priceless knowledge from "Big Name" presenters at deadly earnest sessions. Instead, he confesses to skipping conference panels in order to observe his bartender's skills. In "Lamentations" he and coauthor James Baumlin declare: "Truth is always jacklegged, though it seems static and permanent" (23). Elsewhere Baumlin and Corder elaborate: In West

Texas, jackleg carpentry occurs when "upon completion of a job, the carpenter backs off, surveys the work, and says 'Well, there it is, by God— it ain't much, but it'll hold until we can think of something better'" ("Jackleg" 18). Then Corder and Baumlin reject the demand for definitive answers.

Corder also charms readers by making them laugh. During the 1980s, especially, he laces essays with self-effacing humor, thereby violating the self-important solemnity of academic discourse. One particularly droll essay, "The Rock-Kicking Championship of the Whole World Now and Forevermore," lampoons Americans' obsession with football and, by incorporating biblical references to rock-kicking, burlesques scholars' insistence on citing authorities in their attempts to validate their ethos and justify their claims. Although Corder's easygoing persona and humor create the appearance of accessibility, they mask highly sophisticated argumentation.

IV. ARGUMENT AS INCOMPLETENESS AND FALLIBILITY

Instead of striking an all-knowing scholarly pose, Corder characteristically insists on his own ignorance. He also constantly hedges. At times (for example, in *Uses of Rhetoric*), he qualifies statements so often that a reader must work a bit to grasp the assertions amid the qualifiers. Furthermore, he repeatedly deprecates himself.

Alluding to the final chapter of Samuel Johnson's *Rasselas*, Corder finishes *Uses of Rhetoric* with a self-effacing chapter titled "Certain Maxims and Questions with No Conclusion to be Found." Similarly, "A New Introduction to Psychoanalysis" features what he terms "a diminished ending" (168). Both conclusions satirize scholars' tendency to march readers through a formidable series of arguments to a seemingly inevitable, forceful crescendo.

Similarly, an essay titled "Notes on a Rhetoric of Regret" underwhelms readers accustomed to the usual Argument as Grand Claim. Another title also mocks its author: "From Rhetoric to Grace: Propositions 55-81. . . ." The essay contains no references to Propositions 1-54—an absence that renders its propositional numbering absurd and its author as self-deprecating. "What I Learned at School," Corder's Braddock Award-winning essay, propounds the Ninth Law, Eleventh Law, Eighteenth Law, Twenty-Fifth Law, Twenty-Sixth Law, Twenty-Seventh Law, and Thirty-Second Law of Composition—not any other ones, such as the First Law through the Eighth Law (47-49). He asks readers to provide the missing Laws of Composition. But if the laws are laws, why does he not know all of them already? Obviously, the readers' laws aren't laws, and his laws aren't either. His numbering ridicules both himself and textbooks that purvey writing as enslavement to rules.

"Learning the Text" records Corder's failure to understand his own mother. "Hunting for *Ethos*" designates an earlier essay as "mostly naive" (299). "Lonesomeness" presents an innocent person from West Texas attempting to wrestle the mighty giant of European postmodernism. Misremembering his childhood constitutes the central theme of his *Chronicle of a Small Town*. In "The Heroes Have Gone from the Grocery Store," he misrecalls an image on a box of cereal. Portions of "Heroes" and two other essays detail writing produced about Cheerios for his first-year students ("Occasion"; "Hoping"). Another essay discusses a similar effort for his students on the subject of fountain pens ("Fountain Pens"). One wonders: Why would anyone tackle such unprepossessing subjects at all? He satirizes his own attempt to squeeze meaning from triviality.

During the 1980s and 1990s, Corder deepens and extends his Argument as Incompleteness and Fallibility by questioning any speaker's or writer's claim to authority.

V. ARGUMENT AS BOLD ASSERTION

Corder's persuasive principles—Argument as Indirection, Argument as Charm, Argument as Mystery, and Argument as Incompleteness—would prove worthless if he had nothing to say. But he propounds four cardinal, unqualified claims:

Rhetoric Is an Architectonic Discipline

Like Kenneth Burke, Corder repeatedly and insistently argues that rhetoric is an extremely capacious enterprise that in his words encompasses "all forms of discourse" ("On the Way" 164; "From Rhetoric to Grace" 17). He maintains that "*all* analysis of writing is rhetorical" ("Rhetorical Analysis" 223); for that reason, "rhetoric belongs at the center of *every* class" (*Uses* 125). He contends that rhetoric can serve as "the model for a new way of being, a paradigm . . . of the examined life" (126). He argues that rhetoric encompasses the study of literature, including largely neglected nonfiction prose.[9] He promotes Invention, Structure, and Style as the best framework for all university curricula (*Uses*). He explains historiography as a rhetoric ("Hunting Lieutenant Chadbourne," *Hunting*). He dismisses psychoanalysis as a science or a religion, insisting that it constitutes another rhetoric ("A New Introduction"). In an abstract he summarizes his argument that "rhetoric might organize literary study, curriculum design, course design, geography, cultural history, and psychoanalytic study" ("From Rhetoric into Other Studies" 95). He further posits a biorhetoric of schizophrenia and other forms of mental illness. Realizing that a biorhetoric would necessitate a rhetoric of cognition, he advocates that, too. He also enumerates biology, physics, and painting as rhetorics ("A New Introduction," "On the Way"). And he explicates cancer as diseased, biorhetorical crossovers of Invention, Structure, and Style ("On Cancer").

Corder embraces fundamental ideas of Aristotle and Cicero (that were repopularized by Edward Corbett, Frank D'Angelo, and Winifred Horner) but stretches them far beyond what Aristotle, Cicero, and other ancients conceived. For Corder the problem is that as definitions of (for example) Invention, Structure, Style, and *oratio* were taught over many centuries, those definitions fossilized in textbooks. Corder reinvigorates Invention, Structure, Style, and *oratio* by stretching them. But stretching also defamiliarizes rhetorical concepts. Corder's reanimating-through-defamiliarizing loosely resembles the project of Kenneth Burke, Hayden White, and Frank D'Angelo to revive four classical tropes from their subsequent petrification and vastly expand their meaning into huge conceptual frameworks that explain, among other things, childhood development.

For Corder, *oratio*, when stretched, becomes, in his words, "drama," "sermon," and "dance" ("Learning the Text" 248) while all discourse (plus painting, the human brain, and the human body) embody Invention, Structure, and Style. In *Chronicle of a Small Town*, Corder resuscitates and problematizes Memory, the fourth canon of classical rhetoric. By claiming that bartending is a rhetoric, Corder mocks his own argument that rhetoric is architectonic ("Some of What").

Ethos Is the Most Important Element in Persuasion

Rhetoricians often prefer to examine logocentric argument, partly because ethos is notoriously difficult to pinpoint (Enos). But throughout his career, Corder explores the complexities of ethos. He contributes an early essay, "Ethical Argument in Amos," and adds "Efficient Ethos in *Shane*." His more extensive "Varieties of Ethical Argument" differentiates five forms of ethos: dramatic, gratifying, functional, efficient, and generative. "Hunting for *Ethos*" and "Hunting Lieutenant Chadbourne" posit and examine postmodern ethos, as (in unusual ways) do the four books he published during his final decade: *Lost in West Texas*, *Chronicle of a Small Town*, *Hunting Lieutenant Chadbourne*, and *Yonder*. Despite distinguishing five forms of ethos, Corder probes ethos much more than he expounds it; like much of his writing, his expansive exploration of ethos forcefully resists summary.

Truth Is Jacklegged

In an early essay, Corder analyzes *Gulliver's Travels* as an investigation of multiple perspectives ("Gulliver"). He argues elsewhere examining a multitude of perspectives is invaluable, for "creation is too rich and varied and copious to be comprehended by a single vision" ("Against" 17).

Corder later proclaims "All our answers are provisional" because "no frame of reference . . . is self-guaranteeing" (*More* 48; "Time" 314). He and Baumlin urge readers to "rejoice" at becoming "provisional self-makers" ("Lamentations" 26). "Truly authoritative discourse," they remark, "should always be in process" ("Jackleg" 20).[10] And "research,"

the coauthors continue, "is, at every stage, *interpretation.*" Failing to
acknowledge uncertainty, they add, means promulgating a "naive
epistemology" ("Opinion" 465, 469).

Because truth is contingent, Corder insists, dogma destroys. Because
rhetoric eludes classifications imposed by English textbooks,
composition formulas hobble students. For those reasons, wise teachers
will refuse to pummel students with rules.[11]

For Tade, Tate, and Corder, presumed certainties hamper the reform
of the undergraduate English major ("For Sale") and, for Corder, the
entire undergraduate curricula and schedule as well (*Uses,* "Proposal").
While professors usually recognize that grading is "arbitrary" because it
"has no ultimate reality" ("Stalking"), many compulsively and
dogmatically overvalue their specializations—a tendency that propels
insidious infighting, which besieged administrators are doomed to
referee.[12] But some administrators and other experts yearn for "a saving
authority" in university education, an authority that does not exist
("Time").

VI. ARGUMENT REQUIRES GENTLENESS AND LOVE

Gargantuan impediments frequently block communication. Corder
defines these obstacles as dogmatic certitude, excessive pride, and
pervasive self-centeredness. In his words, "Each of us lives in a province,
and we measure the world's dimensions by our own" ("Late Word" 27).
Our provincialism frequently prevents us from understanding others and
enables us to offend them with great ease, an ease he details in "When."
To counter self-centeredness, rhetoricians need to spawn and popularize
something resembling Rogerian argument. As he suggests in "From
Rhetoric to Grace" and elaborates in "Argument as Emergence," this
rhetoric should include gentleness, love, and patience, as speakers and
writers learn how, in his words, to "pile time into argumentative
discourse" ("Argument" 31). Corder promotes gentle persuasion through
his principles of Argument as Indirection, Argument as Charm,
Argument as Mystery, and Argument as Incompleteness and Fallibility.

Corder's persuasion profoundly violates the norms of scholarly
writing and defamiliarizes rhetoric. He seems to whisper to readers: You
think that academic writing is neutral and impersonal, but I'll show that
it's deeply personal. You assume that scholars love to display knowledge,
but I'll openly confess ignorance. You believe that academics prize
consistency, but I'll directly and repeatedly contradict myself. You
presume that scholars never narrate, but I relish telling stories. You know
that scholars brandish jargon, but I avoid it. You assume that English
professors worship textbook prescriptions, but I'll dissolve those
strictures. You're sure that you comprehend rhetoric, but I'll vastly
expand Invention, Structure, Style, Memory, ethos, and *oratio.* You

define oxymoron as a mere trope, but I'll demonstrate that it results from electroshock therapy on a depressed patient. You view persuasion as hard-headed logos, but I'll affirm that indirection, gentleness, humor, mystery, and even love are fundamental to persuasion. Why is Corder's system of rhetoric radical, pioneering, subversive, and feminist? Consider the rhetoric that Cynthia Caywood and Gillian Overing decry in their groundbreaking 1987 collection, *Teaching Writing: Pedagogy, Gender, and Equity.* In the dominant, masculinist rhetoric, they explain that

> certain forms of discourse and language are privileged: the expository essay is valued over the exploratory; the argumentative essay set above the autobiographical; the clear evocation of a thesis preferred to a more organic exploration of a topic; the impersonal, rational voice ranked more highly than the intimate, subjective one. The valuing of one form over another requires that the teacher be a judge, imposing a hierarchy of learned aesthetic values, gathered from ideal texts, upon a student text. (xii)

Corder's rhetoric is radically feminist because it generally avoids certainties. Because it affirms multiple perspectives. Because it assails rigidities. Because it undermines hierarchy. Because it invites dialogue. Because it valorizes puzzles. Because it prompts laughter. Because it proposes that rhetorical studies encompass poetry and narration. Because its strategies of indirection—circling, bouncing, building clotheslines, and contradicting one's self—are decidedly rare and important. Because it probes ethos at length, refusing to oversimplify. Because its nonlinearity parallels the nonlinearity of diaries, journals, and letters—forms of writing still devalued in English curricula but often favored by women. Because—like many authors of diaries, journals, and letters—Corder develops a homemade persona to mitigate the complexities of his nonlinearity. Because that persona becomes intimate and confessional in *Yonder.*

While Corder never claims that his rhetoric is feminist, he directly renounces models of masculinity that he encountered as a boy and adolescent. He explains a childhood fascination with the movies of Hopalong Cassidy and other manifestations of a pop culture that promulgate the "Code of the West," which demands that males exhibit "physical strength" while remaining "stoical in the face of pain" and "unweeping in the face of grief" ("World War II" 25-27).[13] According to the Code, the "second best place" to "prove yourself [a man]" was "in athletic competition," and "the best place to prove yourself a man was in war" ("World War II" 26-27). He recalls: "*Life* and *Look* and *Time* and the movies and comic books" failed to mention the slaughter caused by American bombs—an omission that helped socialize boys to obey the Code ("World War II" 25; "Lessons").

Only later, Corder notes, did he "learn that the gallant lads of the RAF were from at least 1942 onward deliberately bombing enemy civilian populations," specifically those in Hamburg, Germany ("From Rhetoric into Other Studies" 103; "Lessons" 19-20). "I regret," he remarks, "that I was ever attracted by war stories, and I regret that I ever thought war to be the appropriate test for manhood" ("World War II" 26). He concludes, "I'm not going to study war anymore. Or competition. Or manhood defined in those old terms" (28).

Unfortunately, not everyone learned. Corder and Baumlin observe: "Sometimes we applaud Dirty Harry and Rambo. Sometimes we erect statues to John Wayne and elect his impersonators to the White House" ("Lamentations" 12). Corder laments that during the Gulf War, "The television news and the speeches and the magazines were just about as evasive as the movie and comic book and magazine versions of World War II" ("World War II" 28.)

In "A New Introduction," Corder interprets gang rape as an extreme expression of a generally violent American culture that stupidly prizes machismo (164-67). His *Contemporary Writing*, a textbook for undergraduates, features a brief, but unmitigated denunciation of the Pentagon's absurdly mechanistic and "chilling" description of war. He sandwiches his anti-Pentagon protest inside two passages that decry "brutal and violent words suggesting total male domination" of women (194-97). Whereas "A New Introduction" explicitly links the Code to horrific violence against women, *Contemporary Writing* implicitly—but powerfully—establishes a similar link as it juxtaposes the reductive, inhuman language of sexism and the reductive, inhuman language of war. The terrifying rhetoric of American violence can reach into the English classroom: "Arrogance, ignorance, and dogma . . . pull the trigger, drop the bomb, and teach the student/scholar to be silent" ("Time" 318).

Through these denunciations Corder obviously calls for an alternative to the Code of Hopalong Cassidy and *Life* magazine. Corder asks: "[H]ow do I revise that rhetoric that was and make a new rhetoric that lets me live without my heroes?" ("From Rhetoric into Other Studies" 103). By the time Corder posed that question in 1993, he might have answered it by citing practically everything he ever wrote, for he had already spent virtually his entire career developing a rhetoric that eschews arrogance. Without exactly saying so, in over five dozen essays and books issued over several decades, Corder undermines the argument of domination and enacts an extraordinarily valuable, radical, feminist rhetoric.

NOTES

[1]Corder wrote the third, fourth, fifth, and sixth editions of *Handbook of Current English* and, together with John Ruszkiewicz, coauthored its seventh and eighth editions.

[2]These manuscripts are "Places in the Mind;" "Scrapbook;" and "Rhetoric, Remnants, and Regrets."

[3]I am including *Lost in West Texas*, *Chronicle of a Small Town*, *Yonder*, and *Hunting Lieutenant Chadbourne*.

[4]Although Harris's book is brief, one might think that his extensive discussion of the pedagogy and politics of personal "voice" would include Corder. But it doesn't. Harris only mentions Corder in an endnote in which he mistakenly includes Corder's *Finding a Voice* in a list of ten composition textbooks. *Finding a Voice* is a literary anthology, not a composition textbook.

[5]For the tendency to disappear, see "Losing Out" 99, "Turnings" 111, and *Yonder*.

[6]I don't claim that Corder originated his argumentative principles and strategies, merely that his principles and strategies were highly unusual in scholarly writing during the period of his career. A fair amount of his argumentation overlaps that found, for example, in eighteenth-century British essays and/or twentieth-century American nonfiction—both of which he enjoyed reading and occasionally wrote about.

[7]See, for example, "Against a Mournful Wind," "Some Things," "Going Home," "Episodes," "I Can't Get Away from Hoppy," "Late Word from the Provinces," and *Lost in West Texas*.

[8]See "Late Word" 25, "Humanism" 191, "Going Home" 33, and "From Rhetoric into Other Studies" 105. Corder renders one of these "theological" statements about the Double Mountains even more cryptic by adding a self-contradictory comment: "Refraction turns them blue, though they're not" ("From Rhetoric into Other Studies" 105).

[9]See "Asking," "Rhetoric and Literary Study," and *Uses*.

[10]See also "Humanism."

[11]See "Outhouses," "What," "Studying," and "Academic Jargon."

[12]See "Stalking," "Caught," "From an Undisclosed," and "Tribal."

[13]See also "I Can't Get Away from Hoppy" and "Rhetoric, Remnants, and Regrets."

WORKS CITED

Baumlin, James S., and Jim W. Corder. "Jackleg Carpentry and the Fall from Freedom to Authority in Writing." *Freshman English News* 18 (Spring 1990): 18-25.

Burke, Kenneth. *Grammar of Motives*. 1945. Berkeley: U of California P, 1969.

Caywood, Cynthia, and Gillian Overing. "Introduction." *Teaching Writing: Pedagogy, Gender, and Equity*. Ed. Cynthia Caywood and Gillian Overing. Albany: State U of New York P, 1987.

Corbett, Edward P. J. *Classical Rhetoric for the Modern Student*. New York: Oxford UP, 1965.

Corder, Jim W. "Academic Jargon and Soul-Searching Drivel." *Rhetoric Review* 9 (Spring 1991): 314-26.

—. "Against a Mournful Wind." *New Mexico Humanities Review* 4 (Spring 1981): 13-17.

—. "A New Introduction to Psychoanalysis, Taken as a Version of Modern Rhetoric." *PRE/TEXT* 5 (Fall/Winter 1984): 137-71.

—. "A Proposal for a New Kind of Liberal Arts Core Curriculum, Conceptual, Not Canonical." *Perspectives* 16 (Winter 1986): 27-31.

—. "Argument as Emergence, Rhetoric as Love." *Rhetoric Review* 4 (Sept. 1985): 16-32. Rpt. in *Professing the New Rhetorics: A Sourcebook*. Ed. Theresa Enos and Stuart C. Brown. Englewood Cliffs, NJ: Prentice-Hall, 1994. 412-28.

—. "Asking for a Text and Trying to Learn It." *Encountering Student Texts*. Ed. Bruce Lawson, Susan Sterr Ryan, and W. Ross Winterowd. Urbana, IL: NCTE, 1989. 89-98.

—. "At Last Report I Was Still Here." *The Subject is Writing*. Ed. Wendy Bishop. Portsmouth, NH: Boynton/Cook, 1993. 261-66.

—. "Caught in the Middle." *Liberal Education* 68 (Spring 1982): 69-74.

—. *Chronicle of a Small Town*. College Station: Texas A&M UP, 1989.

—. *Contemporary Writing: Process and Practice*. Glenview, IL: Scott, Foresman, 1979.

—. "Efficient Ethos in *Shane*." *Communication Quarterly* 25 (Fall 1977): 28-31.

—. "Episodes in the Life of a West Texas Materialist." *New Mexico Humanities Review* 1 (Jan. 1978): 29-35.

—. "Ethical Argument in Amos." *The Cresset* 35 (Jan. 1972): 6-9.

—. "Fountain Pens and First Lessons." Unpublished essay, 1994. Author's files.

—. "From Rhetoric to Grace: Propositions 55-81 about Rhetoric, Propositions 1-54 and 82 et seq. Being as Yet Unstated; Or Getting from the Classroom to the World." *Rhetoric Society Quarterly* 14.1/2 (1984): 15-29.

—. "From Rhetoric into Other Studies." *Defining the New Rhetorics*. Ed. Theresa Enos and Stuart C. Brown. Newbury Park, CA: Sage, 1993. 95-108.

—. "From an Undisclosed Past into an Unknown Future." *Liberal Education* 68 (Spring 1982): 75-78.

—. "Going Home." *New Mexico Humanities Review* 5 (Summer 1982): 25-33.

—. "Gulliver in England." *College English* 23 (Oct. 1961): 98-103.

—. *Handbook of Current English*. 6th ed. Glenview, IL: Scott, Foresman, 1981.

—. "The Heroes Have Gone from the Grocery Store." *Arete: The Journal of Sports Literature* 5 (Fall 1987): 73-78.

—. "Hoping for Essays." *Literary Nonfiction*. Ed. Chris Anderson. Carbondale: Southern Illinois UP, 1989. 301-14.

—. "Humanism Isn't a Dirty Word." In Enos "Voice as Echo of Delivery." 190-92.

—. "Hunting for *Ethos* Where They Say It Can't Be Found." *Rhetoric Review* 7 (Spring 1989): 299-316.

—. *Hunting Lieutenant Chadbourne.* Athens: U of Georgia P, 1993.

—. "Hunting Lieutenant Chadbourne." *Ethos: New Essays in Rhetorical and Critical Theory.* Ed. James Baumlin and Tita Baumlin. Dallas: Southern Methodist UP, 1994. 343-65.

—. "I Can't Get Away from Hoppy." *New Mexico Humanities Review* 33 (1990): 107-13.

—. "Late Word from the Provinces." *New Mexico Humanities Review* 2 (Summer 1979): 24-27.

—. "Learning the Text: Little Notes on Interpretation: Bloom, *Topoi,* and the *Oratio.*" *College English* 48 (March 1986): 243-48.

—. "Lessons Learned, Lessons Lost." *Georgia Review* 46 (Spring 1992): 15-28.

—. "Losing Out." *Diversity: A Journal of Multicultural Issues* 1 (1993): 97-100.

—. *Lost in West Texas.* College Station: Texas A&M UP, 1988.

—. *More Than a Century.* Fort Worth: Texas Christian UP, 1973.

—. "Notes on a Rhetoric of Regret." *Composition Studies/Freshman English News* 23 (Spring 1995): 94-105.

—. "Occasion and Need in Writing: An Annotated Essay." *Freshman English News* 16 (Winter/Spring 1988): 3-10.

—. "On Cancer and Freshman Composition, Or the Use of Rhetorical Language in the Description of Oncogenetic Behavior." *CEA Critic* 45.1 (1982): 1-9.

—. "On the Way, Perhaps, to a New Rhetoric, But Not There Yet, and If We Do Get There, There Won't Be There Anymore." *College English* 47 (Feb. 1985): 162-70.

—. "Outhouses, Weather Changes, and the Return to the Basics in English Education." *College English* 38 (Jan. 1977): 474-82.

—. "Places in the Mind." Unpublished manuscript. Author's files.

—. "Religion and Meaning in *Religio Laici.*" *PMLA* 82 (May 1967): 245-50.

—. "Rhetoric and Literary Study." *College Composition and Communication* 32 (Feb. 1981): 13-20.

—. "Rhetoric, Remnants, and Regrets." Unpublished manuscript. Author's files.

—. *Rhetoric: A Text-Reader on Language and Its Uses.* New York: Random, 1965.

—. "Rhetorical Analysis of Writing." *Teaching Composition: Ten Bibliographical Essays.* Fort Worth: Texas Christian UP, 1976. 223-40.

—. "The Rock-Kicking Championship of the Whole World Now and Forevermore." *Arete: The Journal of Sports Literature* 4 (Spring 1987): 1-6.

—. "Scrapbook." Unpublished manuscript. Author's files.

—. "Some Things Change, and Some Things Don't." *New Mexico Humanities Review* 7 (Summer 1984): 19-21.

—. "Some of What I Learned at a Rhetoric Conference." *Freshman English News* 15 (Spring 1986): 11-12.
—. "Stalking the Wild Grade Inflator." *Liberal Education* 69 (Summer 1983): 173-77.
—. "Studying Rhetoric and Teaching School." *Rhetoric Review* 1 (Sept. 1982): 4-36.
—. "The Time the Calvary Didn't Come, or the Quest for a Saving Authority in Recent Studies in Higher Education." *Liberal Education* 71.4 (1985): 305-19.
—. "Tribal Virtues." *Liberal Education* 69 (Summer 1983): 179-82.
—. "Turnings." *Teaching Composition in the 90s: Sites of Contention.* Ed. Christina Russell and Robert MacDonald. New York: HarperCollins, 1994. 105-17.
—. *Uses of Rhetoric.* New York: Lippincott, 1971.
—. "Varieties of Ethical Argument, with Some Account of the Significance of *Ethos* in the Teaching of Composition." *Freshman English News* 6 (Winter 1978): 1-23.
—. "What I Learned at School." *College Composition and Communication* 26 (Dec. 1975): 330-34. Rpt. in *A Writing Teacher's Sourcebook.* Ed. Gary Tate and Edward P. J. Corbett. New York: Oxford UP, 1983. 163-69. Rpt. in *On Writing Research: The Braddock Essays, 1975-1998.* Ed. Lisa Ede. Boston: Bedford/St. Martin's, 1999. 43-50.
—. "When (Do I/Shall I/May I/Must I/Is It Appropriate for me to) (Say No to/Deny/Resist/Repudiate/Attack/Alter) Any (Poem/Poet/Other Piece of the World) for My Sake?" *Rhetoric Society Quarterly* 18.1 (1988): 49-68.
—. "World War II on Cleckler Street." *Collective Heart: Texans in World War II.* Ed. Joyce Gibson Roach. Austin, TX: Eakin P, 1996. 18-28.
—. *Yonder: Life on the Far Side of Change.* Athens: U of Georgia P, 1992.
—. "You Operationalize—I'll Plug Away." *Liberal Education* 66 (Winter 1980): 440-45.
—, ed. *Finding a Voice.* Glenview, IL: Scott, Foresman, 1973.
Corder, Jim W., and James S. Baumlin. "Lamentations for—and Hopes against—Authority in Education." *Educational Theory* 38 (Winter 1988): 11-26.
—. "Lonesomeness in English Studies." *ADE Bulletin* 85 (Winter 1986): 36-39.
—. "Opinion Is, of Course, Bad; Research, on the Other Hand, Is Quite Good: The Tyranny (Or Is It Myth?) of Methodology." *Journal of Higher Education* 58 (July/August 1987): 463-70.
Corder, Jim W., and John Ruszkiewicz. *Handbook of Current English.* 8th ed. Glenview, IL: Scott, Foresman, 1989.
Corder, Jim W., and Lyle Kendall. *A College Rhetoric.* New York: Random, 1962.
D'Angelo, Frank. "The Four Master Tropes: Analogues of Development." *Rhetoric Review* 11 (Fall 1992): 91-110.

—. *Process and Thought in Composition*. Cambridge, MA: Winthrop, 1980.

—. "Prolegomena to a Rhetoric of Tropes." *Rhetoric Review* 6 (Fall 1987): 32-40.

Enos, Theresa. "Voice as Echo of Delivery, Ethos as Transforming Process." *Composition in Context*. Ed. W. Ross Winterowd and Vincent Gillespie. Carbondale: Southern Illinois UP,1994. 180-95.

Harris, Joseph. *A Teaching Subject: Composition Since 1966*. Upper Saddle River, NJ: Prentice, 1997.

Ortega y Gasset, Jose. *What Is Philosophy?* Trans. Mildred Adams. New York: Norton, 1960.

Roundy, Jack. "Crafting Fact: Formal Devices in the Prose of John McPhee." *Literary Nonfiction*. Ed. Chris Anderson. Southern Illinois UP, 1989. 70-92.

Tade, George, Gary Tate, and Jim W. Corder. "For Sale, Lease, or Rent: A Curriculum for an Undergraduate Program in Rhetoric." *College Composition and Communication* 26 (Feb. 1975): 20-24.

White, Hayden. *Tropics of Discourse*. Baltimore: Johns Hopkins UP, 1978.

4

The *Uses of Rhetoric*

W. Ross Winterowd
Professor Emeritus, University of Southern California

᭞As Keith D. Miller points out in his "re-review," *Uses of Rhetoric* is pretty much a forgotten book—even a book that was never widely known. It was seldom, if ever, cited in the professional literature, and copies are now hard to come by. In the vast University of California collections, there are copies only at Davis and San Diego. There is no copy in the University of Southern California libraries.

The *Uses of Rhetoric* is an exceptionally fine book, and it should have had a significant impact on the field of composition-rhetoric. But it didn't. This chapter explains why *Uses* failed and why it should have been important.

The 1960s and 1970s were a time of foment and change in composition-rhetoric. Consider the books that were appearing: Corbett, *Classical Rhetoric for the Modern Student* (1965); Macrorie, *Telling Writing* (1970); Young, Becker, and Pike, *Rhetoric: Discovery and Change* (1970); Elbow, *Writing Without Teachers* (1973); Winterowd, *The Contemporary Writer* (1975); Berthoff, *Forming, Thinking, Writing* (1978). All of these textbooks are intended or unintended counterstatements to the current-traditionalism of such as *The Harbrace College Handbook* (Hodges) (1946 et seq.), and they fall into two clear-cut categories: the rhetorical (Corbett; Young, Becker, and Pike; Winterowd) and (for lack of a better term) the neo-Romantic (Macrorie; Elbow; Berthoff).

Whatever Jim Corder might have intended, *Uses of Rhetoric* is in effect a theoretical and philosophical background for the rhetorical counterstatement to current-traditionalism. And that certainly is one reason for the obscurity into which *Uses* fell. My argument is this: Because of the tacit epistemologies that undergirded them and the values that they expressed, the neo-Romantic textbooks were accepted without question in the English-department establishment, of which composition was, and still to a large extent is, a part; on the other hand, the epistemology and values embodied in the rhetorical texts were so foreign to the English-department establishment that only evolution (the gradual change that rehabilitated literature as discourse about the human condition, not as autonomous structures of meaning) followed by

revolutions (among others, the rhetorical turn that deconstruction
brought about; the concurrent breakdown of "English" as the study of a
canon; the job crisis for new PhDs in English) could move the minions of
sweetness and light. [1] In other words, the establishment was not yet ready
for *Uses*. It may seem odd that I use the term *establishment* in referring
to those folk whose professional commitment even in part is to
composition, but just as surely as junior executives buy into the corporate
culture of which they have become a part, so the heterogeneous bunch
that made up the composition faculty (at state colleges, many tenured
professors and some part-timers; at major universities, part-timers and
graduate students) absorbed and promulgated the values of "the" English
department. Those values, attitudes, and practices needed no defense;
they were simply the way things were. The humanities as defined in the
English department (that is, "imaginative" literature) and the hierarchy
of values assumed in the English department (that is, imagination, a
faculty infinitely superior to invention, the mechanical process of finding
means of persuasion in regard to any subject whatever) were the pinnacle
of the liberal education and the very basis of social grace. Without those
humanities one could scarcely be human.

The heart of *Uses* is its explanation of invention and its argument
that invention should form the basis for the humanities. [2] Well aware of
the vagaries of publishing, the lottery of reviews, the blind luck of finding
readers—in other words, totally cognizant of how iffy the book business
is—I nonetheless strongly suspect that *Uses* was the right book at the
wrong time; its statement was so foreign to the establishment that the
book was doomed. (During the 1990s a number of books defending the
neo-Romantic stance in composition appeared, among them Knoblauch
and Brannon, *Critical Teaching and the Idea of Literacy* (1993); Gradin,
Romancing Rhetorics (1995); Roskelly and Ronald, *Reason to Believe*
(1998). The interesting aspect of these treatises at this point in history is
not what they say but what they imply about composition: Since they
passionately defend the neo-Romantic tradition, that tradition must be in
danger, or perhaps it is moribund. These counterstatements to *Uses* are a
happy omen for rhetoric.)

Jim Corder, then, was defending the indefensible. Two books in the
rhetorical tradition did prevail, Corbett's *Classical Rhetoric for the
Modern Student* and Young, Becker, and Pike's *Rhetoric: Discovery and
Change*, but they were textbooks that were adopted by either true
believers or by the naive who didn't know what they were getting into.
These two books helped provide rhetoric with the staying power it
needed.

Lurking in the background of American romanticism are, among
others, Coleridge, Wordsworth, and Emerson, who set forth the doctrines
of creativity and imagination that ran directly counter to invention. In a
new twist on the old faculty psychology, Coleridge dreamed up two
imaginations, the primary and the secondary, and the primary
Imagination became a god-like faculty, creating original works (that is,
poetry, fiction) while the secondary Imagination was relegated to a

servile position in the scheme of things, fashioning such utile, practical works as essays, letters, biographies, histories, and so on. In regard to Wordsworth, I have argued elsewhere that

> [f]rom *The Prelude* emerge three foundational principles of Romantic invention: first, the poet's only creative resources are imagination and intuition; second, the subject matter is himself; and, third, empiricism misleads and even corrupts. Thus all the resources for invention are "in here": the *method* (imagination and intuition) and the subject matter (the self); any empirical reliance on, or nominalistic belief in, what is "out there" can only mislead. (*The English Department* 119)

Emerson's romantic epistemology is more complex than that of Coleridge and Wordsworth, but the case against the great and ubiquitous transcendentalist boils down to this passage from "Self-Reliance": "To believe your own thoughts, to believe that what is true for you in your private heart is true for all men—that is genius."

Against this phalanx of heavyweights in the humanities comes Jim Corder, with "The arts of invention are occasions for exploration, plunging into experience, testing all possibilities" (*Uses* 111). The contrast is striking.

Some of *Uses* is—what term should I use?—humdrum. But this plodding through familiar territory is interesting in itself and was necessary for Corder's (thwarted) purpose of rehabilitating rhetoric. That is to say, what would not have been news to, for instance, E. P. J. Corbett or Richard Young would have been revelation to English-department humanists. Thus, Corder must overcome "Misconceptions of Rhetoric" (4-8) and explain the traditional departments, the concept of topics, and the *progymnasmata*. For example, "Invention is the name given to various meditative and investigative arts and processes (sometimes lumped today under the heading *prewriting*) necessary for the discovery of what might be said and what should be said" (*Uses* 22). In fact, chapter 2, "Shapes and Directions in Rhetorical Study," is an excellent beginners' introduction to rhetoric. The paradox is just this: Much of the rest of the book demands a good deal more than a beginner's knowledge if one is to understand the subtlety and complexity of Corder's argument. He was in a double bind, writing for an audience that needed basic education, but advancing an argument foreign to that audience's values and beyond what a reader with only the basics could follow. Not, of course, that *Uses* is arcane or unduly complex; it is a graceful and generous book, but, like all other worthwhile books, one must have some preparation and background to understand it.

Corder argues that "in the culture-view the necessity of invention has been dismissed in favor of spontaneity and simultaneity, expectations out of our silent background that make loud and irresistible, though often unacknowledged claims upon us" (*Uses* 51). And then,

> I must delay to be sure of my point. Our culture has not been
> antagonistic to invention. Invention has not been an issue. The
> reason it has not been an issue is that in our complex culture-
> view it has already been effectively dismissed. Our indis-
> position to invention, that is to say, is part of a much larger
> whole against which invention and its kindred arts are an
> indispensable corrective. "Every philosophy," Alfred North
> Whitehead said, "is tinged with the coloring of some secret
> imaginative background, which never emerges explicitly in its
> train of reasoning." There is that in our "secret, imaginative
> background" that dismisses invention as no issue, and we
> deprive ourselves, as children do, of the pleasures and benefits
> of such sequential pursuits as I hope to describe later,
> participating as we go in the disenthronement of knowledge
> that gives the primary hue to our background. (51)

Taking Wordsworth as his representative Romantic (or representative anecdote), Corder gets right to the heart of the problematic clash between inspiration and invention. Wordsworth's definition of poetry as "the spontaneous overflow of powerful feelings" sets up the conditions for ennobling spontaneity in all discourses or, rather, ennobles the sort of discourses that come about through spontaneity, not deliberation (that is, poetic discourses as opposed to nonpoetic). "The point is, I think, that many of Wordsworth's generation (and much earlier for that matter) had already accepted this vision of our natural plenitude and rightness, even if they did not state it plainly and explicitly" (*Uses* 54). It is not a great leap from the epistemologies of Coleridge, Wordsworth, and Emerson "to the modern text[book] that begins with the assumption that all students have something to say" (54).

Cases in point are easy to come by. Anne Berthoff stresses the making of meaning, words evoking other words endlessly:

> By proceeding philosophically—by reclaiming the imagina-
> tion—those learning to write and to teach writing will discover
> that language itself is the great heuristic. Language enables us
> to make the meanings by whose means we discover further
> meanings. We don't think *about* concepts, we think *with* them.
> (*Reclaiming*, "Preface")

And Jane Tompkins, speaking about one of the "assignments" in her composition class:

> We drove over [to Toys "R" Us®] in three or four cars, spent
> forty-five minutes in the store—the rules were no talking to
> anyone, and you could only buy one thing—and we came back
> to the classroom and wrote for the remainder of the period. For
> next time, everyone finished the writing assignment, to write
> eight pages on what the Toys "R" Us® experience had brought
> up. (138)

Corder's reaction to these examples might have been:

> It should not be surprising to us to see that rhetorics [that is, textbooks] of the nineteenth and twentieth centuries not only abandon invention; they also abandon that preliminary searching for external resources. Save for that solitary research paper so common in freshman English courses, it seems accepted always that internal resources are sufficient. (*Uses* 55)

Corder is asking for reason and deliberation, for argument rather than diatribe, and the basis for his appeal is a decidedly unfashionable view of the human condition. The new vision of humankind's natural rightness and plenitude "replaced an ancient belief in the omnipresence of evil, demanding for man's salvation (here or later) the constant vigilance of study, deliberation, and self-negation" (*Uses* 55). This theistic overtone was out of keeping with the prevailing attitudes of the 1970s and is out of keeping with the prevailing attitudes of the 1990s. One simply doesn't moralize, particularly in the context of theism.

By the middle of the eighteenth century, people had, in the words of Samuel Johnson, been of "a disposition to rely wholly upon unassisted genius and natural sagacity." Corder goes on to explain:

> It has been our malady since; we were quickly and completely convinced of our own infallibility. We were taught to trust ourselves as we learned the catechism of the sentimentalist: man is good, man is right, man cries out of no depths he cannot himself escape. As we catechized, we were nurtured by latitudinarian impulses in the church that fostered sweet benevolence rather than belief, by expansions of scientific knowledge that allowed us to see that the world was not, after all, in a state of decay since the Fall, but rather the continually self-perfecting work of a creating God whose endeavors were ultimately replaced by our own, and by economic changes that brought masses unacquainted with and frequently uninterested in classical and humanist disciplines not out of the gutter, but at least up from their knees. We have come at last to think ourselves good without question, our native wit sufficient without learning, our feelings right without deliberation. (Qtd. in *Uses* 55)

For Corder the loss of invention and of belief were concomitant.

Current-traditional rhetoric was devoid of invention; neo-Romantic rhetoric is devoid of craft, that is, of considerations of disposition and style. One of the many apothegms in *Uses* is this: "[A]s *invention* is an invitation to openness, *disposition* and *style* are invitations to closure" (166). Corder is proposing the corrective to both current-traditional doctrine and pedagogy, which are devoid of invention, and to neo-Romantic doctrine and pedagogy, which are devoid of craft, that is, of disposition and style.3

In the context of controversies over instruction in English as a second language, Ebonics, and "the basics," Corder's recommendations are particularly apt. He says, "All study begins from and eventuates in some linguistic form, some style" (*Uses* 171), and one main purpose of education is to give students the fluency to switch styles, to escape their own native "dialects" and enter into the discourse fields of academia and the learned world. Escape is, of course, a movement outward, not inward; one does not escape into prison. To enter the world of learned discourse, one must have not only the substance but also the forms of that discourse, and the two are inseparable. In learning economics one also learns the styles and forms in which the discourse of economics takes place. Thus, "it seems to me that a graduated, sequential, and flexible set of writings and speakings should be the center of any curriculum—not necessarily to belong to or be supervised by an English faculty, but to belong to and be supervised by us all" (171).

It is not an oversimplification (though it is a truism) to state that current-traditional textbooks were all style and disposition. Adams Sherman Hill's *The Principles of Rhetoric* (1878, 1895) can fairly be taken as the prototype:

> [Rhetoric] does not undertake to furnish a person with something to say; but it does undertake to tell him how best to say that which he has himself provided. "Style," says Coleridge, "is the art of conveying the meaning appropriately and with perspicuity, whatever that meaning may be"; but some meaning there must be: for, "in order to form a good style, the primary rule and condition is, not to attempt to express ourselves in language before we thoroughly know our own meanings." (vi)

In the context of Hill's massively influential textbook, the quotation from Coleridge might well serve as the basis for an extended disquisition on style and invention in writing. On the one hand, Hill uses Coleridge (who was and is, after all, a sacred figure in "the" English department) to justify the current-traditional vision of rhetoric; on the other hand, neo-Romantics, such as Berthoff, for whom Coleridge is a presiding deity, must circumvent the fen in which they would slosh and struggle if they heeded the Master's epistemology: Know what you want to say before you say it. After all, as Berthoff says and Elbow implies, language makes meaning.[4]

From Corder's discussion of first-year composition, one can alembicate[5] a solution to the neo-Romantic language-meaning conundrum and see the meaning-style-disposition relationship. The purpose of the first semester is "to promote gathering, exploration, and experimentation, not ten or however many essays" (*Uses* 175). In other words (though Corder regrettably does not say so specifically), the first semester would lead students to find out about *something*; they would have a purpose and would not be wandering through the library or around campus or up and down the aisles of Toys "R" Us® hoping to run onto something to say. Invention is always associated with a problem and hence with a purpose. Note the sharp contrast between the following two simple statements: "I'm making meaning." "I'm trying to decide whether or not capital punishment is both effective and moral." Purpose generates ideas that are in language or nowhere and that are the raw material of discourse. Style and disposition refine this material as the writer shapes his or her discourse for an audience.

Corder is saying that we can help students learn to be inventive and that we can help them develop their styles and their sense of form. That is what first-year composition and the university-wide writing program are all about. Corder says that a university-wide writing program should include all of the aims of discourse (as outlined by Kinneavy): referential, with its focus on subject matter; persuasive, with its focus on the reader or hearer; expressive, the writer or speaker's "need to discharge his [sic] emotions, to achieve his own individuality, or voice his aspirations"; and literary (*Uses* 172). Thus, we see in *Uses* a proposal for what has come to be called "writing across the curriculum," but with a proviso: "This study and performance of discourse, as Kinneavy says, is a foundation stone for the liberal arts tradition, and, one might add, for its future"(173).

The *Uses of Rhetoric* is a paradox. In its conservatism it was revolutionary. It argued not that we should return to Aristotle, Cicero, and Quintilian (though certainly we should not lose them) but that the future of the humanities lay in the more recent past, in the age before such figures as Akenside and Addison dreamed up their versions of imagination and certainly before the great Romantics solipsized, and thus did away with, invention.

Let Jim Corder have the last words.

> Invention is always greater than disposition and style. We are never in command of all our resources, weaknesses, and strengths. Of this maxim—let it be a last thesis—this book is, I think, sufficient testimony, though not at all in the expected way. Oh, after all, some uses of rhetoric have been shown, I think, and need no special apology. Yet time and again, when the book has approached some sort of perplexing question, the heart, perhaps, of some matter, it has backed off, asked some more questions, and wondered aloud. The reasons for this are varied: ignorance; failure to understand or to appropriate the terms of rhetoric; impatience; perhaps even a kind of repudiation that takes the form of hurrying from the past to the present. And so I offer not a book that definitively identifies the uses of rhetoric, but rather one that ends as it began, with a plea for the recovery and extension of rhetoric, for the discovery of its uses. And that, too, I think needs no apology, for a plea for rhetoric is a plea for life. Invention is always greater than disposition and style. Yet it is endlessly there, copious, varied, confusing, frustrating, and marvelous, for the shaping of our communion with each other. Each man lives in a boundlessly rich inventive universe. I have been in the

presence of the Lords of the Plains. I have sat at the Comanche fire, and a bedraggled and weary scout walks in my tracks. When I waded in a tributary of Duck Creek, I became Duck Creek; my foot's weight shifted rocks and small currents and a million insects and became the system of feeding, living, begetting. When my bent-feather scout friend left the Double Mountain Fork of the Brazos to chase small game into my canyons, he became, all those years ago, the shape of the rock I threw. I am in the past, and the future is in me. We are in the past, and the future is in us. We shall have the future we deserve, and there will be little luck and small chance, only our ignorance of what has been and already is, to account for our failures. We are in the past, some universe of invention, and it generates our future. All things have consequence; let nothing be lost. (*Uses* 208-09)

NOTES

[1]Developed fully in *The English Department: A Personal and Institutional History*, my critique of both romanticism in composition and English departments in general.

[2]In the current-traditional and neo-Romantic views, invention means "stylistics." See Winterowd, *The English Department*, 32-74.

[3]Disposition or form in current-traditional textbooks tended to be algorithmic. The five-paragraph essay is the most notorious example.

[4]"I've spent a lot of time in a debate with myself about whether it's better to work things out in the medium of words or in the medium of ideas and meanings. . . . After some cogitation, I came to decide that both levels are good, but for different purposes: perspective and immersion. Working in ideas gives you perspective, structure, and clarity; working in words gives you fecundity, novelty, richness" ("Cooking" 42-43).

[5]One of Kenneth Burke's favorite terms. I use it here for a purpose that I hope is obvious.

WORKS CITED

Berthoff, Ann E. *Forming, Thinking, Writing: The Composing Imagination*. Rochelle Park, NJ: Hayden, 1978.

—, ed. *Reclaiming the Imagination*. Portsmouth, NH: Boynton/Cook, 1984.

Corbett, Edward P. J. *Classical Rhetoric for the Modern Student*. New York: Oxford UP, 1965.

Corder, Jim W. *Uses of Rhetoric*. Philadelphia: Lippincott, 1971.

Elbow, Peter. "Cooking: The Interaction of Conflicting Elements." *Embracing Contraries: Explorations in Learning and Teaching.* New York: Oxford UP, 1986. 39-53.

—. *Writing Without Teachers.* New York: Oxford UP, 1973.

Gradin, Sherrie L. *Romancing Rhetorics: Social Expressivist Perspectives on the Teaching of Writing.* Portsmouth, NH: Boynton/Cook, 1995.

Hill, Adams Sherman. *The Principles of Rhetoric.* New York: American Book, 1878, 1895.

Hodges, John Cunyus. *Harbrace College Handbook.* New York: Harcourt, 1946.

Kinneavy, James L. *A Theory of Discourse.* Englewood Cliffs, NJ: Prentice, 1971.

Knoblauch, C. H., and Lil Brannon. *Critical Teaching and the Idea of Literacy.* Portsmouth, NH: Boynton/Cook, 1993.

Macrorie, Ken. *Telling Writing.* New York: Hayden, 1970.

Miller, Keith D. Rev. of *Uses of Rhetoric,* by Jim W. Corder. *Rhetoric Review* (Spring 1999): 331-36.

Roskelly, Hephzibah, and Kate Ronald. *Reason to Believe.* Albany: State U of New York P, 1998.

Tompkins, Jane. *A Life in School.* Reading, MA: Perseus, 1996.

Winterowd, W. Ross. *The Contemporary Writer.* New York: Harcourt, 1975.

—. *The English Department: A Personal and Institutional History.* Carbondale: Southern Illinois UP, 1998.

Young, Richard E., Alton Becker, and Kenneth L. Pike. *Rhetoric: Discovery and Change.* New York: Harcourt, 1970.

5
Preaching What He Practices: Jim Corder's Irascible and Articulate Oeuvre

Wendy Bishop
Florida State University

> *Style and identity are symptoms of each other.*
> -Jim W. Corder ("Hunting")

> *We keep trying to tell our souls to the world. We leave tracks.*
> -Jim W. Corder ("Notes")

☞"I believe that I exist," opens one of Jim Corder's essays ("Learning" 243). Looking through prints and reprints of Corder's work, I come across a biographical note with his birth date, 1930 (though the Library of Congress reports 1929). I respond to this data by thinking that a lifetime is too short to realize the ardently expressed desires—to know and be known, as a writer by readers—that are so deeply inscribed in all of Jim Corder's essays. A narrative rhetorician, born between wars. The great war, I imagine, would be his father's story, and WWII would be the story of his teens and young adulthood. Drafted into the army, he spent 1951-52 in Germany. A young man from the small West Texas town of Jayton goes to Heidelberg and returns to Oklahoma for a PhD in restoration and eighteenth-century English literature. He traveled on, in 1958, to Texas Christian University (TCU) where he spent his academic life. Teaching, writing, administering; teaching, writing, engaging in the study of rhetoric. And, in fact, all his scholarly essays invite us to join him in this, his late-adopted but clearly beloved course of study.

THE CORDERIAN APPEAL

I never met Jim Corder, but I feel a strong affinity for his work on several levels.[1] My father was in the Army but born a generation earlier than

Corder, in 1918. I was born in 1953 in occupation Japan while my father fought in the Korean war. I've not spent time in West Texas though I've driven fast and long through its expanses, and I sometimes think I've constructed an imaginary TCU simply from my reading of Jim's work and imaginings of him there. He was published often but is not someone most would consider widely published: That is, he was known and respected but not canonized in reading lists, not a major player— according to documentary evidence—in the professional battles of his years. Nor did he appear to feel obliged to battle it out for professional visibility. His writings read, always, like those he felt moved to write, compelled to write, not those that a professional trajectory commanded or demanded he write. "I yam what I yam," fits, perhaps.

I never found myself assigned Corder in graduate school, nor did I have regular encounters with Corder-writing in journals. Even though his essays appeared in *Rhetoric Review, Composition Studies/Freshman English News, College English,* and *College Composition and Communication,* they did so over a twenty-year span, and he added to this oeuvre several coauthored or coedited textbooks, a volume of literary scholarship, a volume of rhetorical scholarship, and a fine volume of history and biography. Still, after first meeting him—that is, reading him—I found myself listening for his voice and paying it serious attention when I heard it. This is because all his texts ask us to investigate why we are here teaching writing on planet earth (if we are teaching writing, which we probably are if we are reading his essays in the first place) and ask us to consider what we're accountable for in that complicated enterprise.

In the essays I cite in this chapter, Corder takes strong stands, positions rooted in and growing out of a textualized wrestling that he shares with us. He seeks to understand his identity in relationship to his communities; or as he would put it, his rhetoric(s) in relationship with other rhetorics, something he realizes the daily life of English departments, writing classrooms, and professional networking don't often encourage. Personal though his investigations are, he encourages a rhetorical rather than a romantic stance, arguing for individual *and* community and for understanding that we create ourselves vis-à-vis our connections to each other. In this coconstructive stance, he bridges some of the perplexing divides that have created ideological rifts among those who teach writing from purportedly opposing perspectives. He proposes to those on the opposite canyon rims of socially and individually oriented perspectives that there is no canyon, really; or if there is, rhetoric is always the bridge. Built by cultivating an understanding of our multiple, but overlapping, rhetorics. Built by using the rhetorical tools available to us all.

Briefly then, encountering my first essay of Jim's—"Asking for a Student Text and Trying to Learn It"—proved pivotal in my own teaching and thinking about teaching. A simple and profound insight of this essay, wittily discussed, that he attempted to write all assigned class essays with his composition students, provided crucial support in my struggle to connect my identity as writer with my identity as a teacher of writing. And a wonderfully unexpected bonus: I also enjoyed his essay. I still do.

CORDERIAN ARGUMENTS

Jim's essays practice what they preach *and* preach what they practice. He offers instructive, sometimes irascible, wry, serious discussions of big issues in prose that are both everyday and elegant, playful and artful. His changes of register call me to attention, and then his carefully deployed irony entertains me. "I reckon that texts do exist" ("Asking" 89), he claims. And in another essay explains: "I taught the first course called a rhetoric course at the university where I work. Actually, I called it 'Rhetoric and Chocolate.' In 1966, I didn't think anyone would enroll if I just called it 'rhetoric'" ("Notes" 96). Not only down to earth and humorous but also scholarly and thoughtful in the best sense: reflective. Often, Corder's work reminds me of Peter Elbow's in that he regularly investigates multiple positions, takes both doubting and believing positions; a particularly strong example of this can be found in "Argument as Emergence, Rhetoric as Love," but similar moves occur throughout his texts. While Corder is a great doubter—often this lingers as the irascible side of his textual character—he is also a great believer— in the power of reenvisioning rhetoric as a way to help us join our thinking and writing worlds together, each and every teacher-writer-scholar to each and every teacher-writer-scholar—that is, rhetorician to rhetorician. And, perhaps more important, teacher to student and student to teacher.

Corder's interest in argumentation is (always, already) grounded in rhetoric, the study of which he represents as part of his lifelong journey of understanding. His essays regularly interrogate the historic art: investigating its workings, asking why it has continuing importance, for him and for his students, and why it might make a fine gift for his colleagues. "Argument, then," he argues, "is not something we *make* outside ourselves; argument is what we are. Each of us is an argument" ("Argument" 415).[2] He asks us to put our arguments, our selves, in dialogue together because he believes that is the only way we have of validating and understanding ourselves.

Because he views argument as expansive and generative (and human), in making one, Corder considers a variety of positions by imagining himself in them and by considering his own position in tandem (not simply in opposition) to others: as when a shopper goes down a rack of clothes holding up this shirt and that, trying on without quite trying on, looking for likely fits before getting serious about a few and deciding, perhaps, on the best one(s). And he asked us, often, to consider the unconsidered along with him, as he does here:

> And, after all, how many texts are there on the desk when our
> students turn in a set of papers, and how many of us are
> present? I think there is a text that each wanted to write. I
> think there is a text that each thought he or she wrote. I think
> there is a text that each did write and turn in. That's three, but
> not all. There is the text we hoped they would write (ours).
> There is the text we hoped they would write (theirs). There is
> the text we try to read. That's six, and no doubt there are other
> permutations.[3] ("Asking" 93)

This, like so many of the quotations I share in this essay, strike me as
"classic Corder": investigative, humorous, accurate, and thought
provoking. And the questions he asks are also classic questions, in that
they have long occupied philosophers, rhetoricians, writers. Key for
Corder are the following issues that I've teased out from the interrelated
congresses of thought that comprise his work.

Foremost, Corder asks, If I believe I exist, can I signal that sentience
in the texts I write? More simply: Is there a writer in this text? Or, more
complicatedly, How does postmodern theory propose to deal with this,
my body, and the ethos I develop and attempt as a writer to inscribe as I
make the jump from here to there? Jim wanted to discover—to assert
really—that there was a sound, sensible, benevolent character of high
morals—an ethical being—behind a text of the same sort: that he could
embody his literal character in his text and that, conversely, his text
could be composed in a manner that would well and reasonably
represent his character.

> I assumed that ethos does not always work in the same way
> and proposed five rough and overlapping kinds—dramatic
> ethos, gratifying ethos, functional ethos, efficient ethos, and
> generative ethos.
>   I wanted to believe, you see,[4] that ethos was real and in the
> text, that it was there and I could find it. ("Hunting" 300)

Throughout his essays Jim asserts his existence by sharing family history
and by investigating ethos through attempts at enactment. When I read
his work, I can feel him working to construct his sense of self on the page,
in his texts. This desire to live on through his writing was, in part, what
led me to claim that he did not live long enough to investigate all these
issues that he loved to wrestle with (because there is no end to such
investigations), but perhaps I was wrong; for I feel that he accomplished
his aims to an astonishing degree. Jim Corder is the text he wrote,
leaving traces of a self that a reader can evoke, an evoked self that
resonates. These traces are strong in his work and can be identified and
celebrated through a stylistic analysis of the sort I'll undertake before this
chapter ends.

Not unaware of or disengaged from literary traditions and
contemporary critical theory, the writer in Jim Corder came down on the

side of existence, manifest in his physical body, evident in his textual tracks:

> And yet I am still here, if not on the page, still here and responsible for what is on the page. We are all still here. Turning, it may be, into nothing, spinning, twisting, drifting into oblivion, we are still here, responsible for ourselves, though doubting ourselves, potentially noble, permanently errant. We keep trying to tell our souls to the world. We leave tracks.[5] ("Notes" 97)

He makes this assertion because he has investigated and believes in both the social *and* the personal, the individual *and* the tribe. In his interested runarounds with rhetoric, he comes to view it as process, the process of individual participation in, with, against the tradition. This is how he describes that tradition to freshman writers:

> All of us have been, whether knowingly or not, whether directly or not, part of a rhetorical tradition that has been with us for 2,500 years. This tradition allowed and encouraged us to believe that we could achieve identity not just in what we do, but also in what we say and write, and that others could know us in what we say and write, and that they could tell themselves back to us with their words. ("At Last" 262)

It is clear from reading Jim's essays that he doesn't come to this conclusion lightly. He does worry about his own romance with individuality, but he works through the arguments, coming back always to the equation of "I am therefore I am":

> If I slander someone in what I write, interpreting readers won't go with me to court and sit beside me as codefendants. I'll be alone, and I'll be held accountable. If I lie in what I write, the social group that constructed me won't sit beside me to share the judgment against me. I'll be alone, and I'll be held accountable. I am responsible for what I write. No one else is. Just me.[6] ("At Last" 265)

A celebrant of history, biography, autobiography, Corder is well aware of the social. In fact, I believe it is his fierce sense of history—that the individual is a small spot on the screen of time—that makes him believe so strongly that by writing, through writing, by inscribing and accumulating textual traces, individuals can and do matter. His essays are haunted by memories of his parents, now gone and clearly missed; of the losses of wars, now over yet not forgotten; of small Texas towns, no longer inhabited but once contributing to the richness of American life. It comes down to belief, sometimes, oftentimes. Because of this, Jim Corder would extend to his writing students the same opportunities: to believe as much as they doubt, to write personally *and* academically, to participate in commodious views of writing and rhetoric. His pledge of allegiance is this:

> I do still believe, however, that both academic writing and personal writing in all their diversity, with whatever purity they can attain or with whatever impurity they must reveal, ought to be part of our knowledge and practice.[7] ("Academic" 314)

And it is in this essay, "Academic Jargon and Soul-Searching Drivel," that Corder weighs in on the most hotly contested composition debate of his time (of our time?). The basis of his argument would appear to be a rephrasing of the golden rule: Shouldn't we do unto our students as we do unto ourselves? Shouldn't we preach what we practice? In a similar manner, he asks writing-teacher-readers to write the essays they assign to their students. He asks us to consider how our students might benefit from gaining access to the diverse forms and styles of writing that classical rhetoric prescribes and that we continue to investigate as professional writers. He asks for plenitude: a word, like many of the words he uses, not often proffered in professional venues. At the heart of this and other arguments comes, again, a look at, nod to, awareness of history, of time. Rhetorical exploration in a Corderian universe requires meditation, space, and time (see "Argument"). Time is what writing teachers' lives and writing students' lives often have too little of, or too often that little is offered up within the artificial constraints of class period, term, academic year.

Regularly, Corder describes himself as a slow learner; I find him slow in the best sense: ruminative and reflective. He wanted to learn what rhetoric meant: again to him as an individual, to the profession, to his students. Rhetoric in his understanding reigns as both product *and* process, location *and* destination, place to stand *and* place to leave from on an important, time-intensive speculative journey that returns us to a newly familiar place. We must make that journey to see that others are on similar journeys:

> We are always in a rhetoric. We may see those others in rhetorics not our own; if we do, they are likely to seem whimsical, odd, uninformed, selfish, wrong, mad, even alien. Sometimes, of course, we don't see them at all—they are outside our normality, beyond or beneath notice; they don't occur as humans. Often as not, we don't see our own rhetoric; it is already normality, already truth, already the way to see existence. When we remark, as we have become accustomed to remark, that all discourse is ideological, we probably exclude our own. It is the truth, against which ideological discourses can be detected and measured.[8] ("From Rhetoric" 98)

Jim Corder's arguments are grounded in this, his ongoing investigation of rhetoric. And belief in it. Rhetoric becomes metaphor. He considers at one point if schemes and tropes don't really extend from text to life:

> All of the texts of rhetoric make a text of rhetoric, which we render in our various ways and which I am slow to learn. . . . Other schemes—anaphora, for example, and epistrophe and epanalepsis and anadiplosis—retain their traditional definitions and usages, to be sure, but are also structural and stylistic qualities extending beyond sentences to paragraphs, verses, acts, chapters, *lives*.[9] ("Learning" 245; emphasis added)

Personification, a text with a real body. Jim Corder essays are performative. Textually, he is not certain. He wavers and he explores. He becomes more certain. Loudly so. And then softens and goes on again. His favorite name for himself is not rhetorician but a once-used epithet, Professor Fog ("Notes" 97). Yet there's also a fox in the fog, for he uses this commodious doubting and believing to guide his readers to an intended end (he has a rhetorical purpose)—that of considering individuality in an age of the tribe. Asking "How can we expect another to change when we are ourselves that other's contending narrative?" ("Argument" 416), he assures us that this will be accomplished only through cherishing, acceptance, empathy—love! His journey is toward you, requesting, if you don't agree with him, that you at least accept him. In some ways Jim Corder's essays ask for nothing more than to let him join you because rhetoric is life practice:

> We arguers can learn the lessons that rhetoric itself wants to teach us. By its nature, invention asks us to open ourselves to the richness of creation, to plumb its depths, search its expanses, and track its chronologies. But the moment we speak (or write), we are no longer open; we have chosen, whether deliberately or not, and so have closed ourselves off from some possibilities. Invention wants openness; structure and style demand closure. We are asked to be perpetually open and always closing. If we stand open, we cannot speak or act: if we stand closed, we have succumbed to dogma and rigidity.[10] ("Argument" 425)

Writing teacher and writer. Rhetorician and Professor Fog. Individual and part of the tribe. Both at the macro level—that of his arguments—but also at the micro level—that of his style and sentences. I'm impressed: with the sophisticated fog of Jim Corder's arguments and the consistent ethos of his texts, but also with the way his study of rhetoric seems echoed and amplified by his stylistic undertakings—his moves. They are diverse, linking trope and scheme to lived life.

CORDERIAN STYLE

It seems to me that Jim Corder's prose enacts his arguments, a crucial strategy for the man who—postmodern theories of authorship notwithstanding (which, by the way, he pretty carefully considered)—was always looking for himself: trusting that he—that any writer—would be able to find self in a sentence. "Style and identity are symptoms of each other," he claims confidently, and continues on with Corderian acerbity: "I'm in here. Shall I provide documentation? Are telephone numbers needed? Letters of reference? Will the testimony of a wife, librarian, and bartender suffice?" ("Hunting" 314).

Jim Corder announces himself loudly in his essay titles—just read down the list of Works Cited—and in his general prose style. Neither wife, librarian, or bartender, I'm eager, still, to illustrate how his prose is provocative and of a professional piece (in the way I'd like to redefine professional) because I believe that his life in letters provides a model for how any like-minded scholar can develop a line of thought, a voice—if we want to call it that—that matters. This can be done without relentless publication. It can be accomplished by writing out of belief, interest, and urgency (the personal-professional kind) instead of (only or primarily) institutional exigency. The former remains in mind after the writer has gone. The latter fills our files until we retire and it disappears.

You may have already caught a whiff of this in earlier quotes, but I'm always engaged with the way Corder invites West Texas into his texts. He cusses and praises, mixes the practical with the erudite, country with city, town with gown. In doing so he argues for his beloved diverse discourse: intentionally and skillfully playing the full spectrum of language registers. He also includes keywords that are not common in academic discussions or in professional-writing communities that tend to value the author-evacuated, evenhanded temperance of the sciences. Some examples:

Jim Corder makes a point of including folksy argumentation and instances of plain-speaking (I'm reminded of Albert the Alligator in *Pogo* when I read "headbones" in the quotation below, and I'm willing to hazard the allusion is intentional):

> [W]e can also remember that there are many laboratories for research into the teaching of writing. One of them is inside our own headbones. ("Academic" 326)

> I have often begun sentences with *and* or *but*. Mrs. Solon, my eleventh-grade English teacher, told me not to do that, but I have done it anyway because, in the first instance, I reckon that the previous sentence doesn't say enough, and, in the second instance, I reckon that it says too much. ("Notes" 94)

> And if someone tells us that YOU MUST KNOW AND ADJUST
> TO YOUR AUDIENCE, we'll know that sometimes we have to
> go the hell on without an audience. ("Academic" 326)

> I sometimes privately think of myself as a writer—it's a helluva
> lot better than thinking of yourself as a dean, or court scribe, or
> fool. ("Hunting" 308)

We're used to talking about miracles in church or in popular songs, not in academic essays. During scholarly discussions, as a rule, love and plenitude are proscribed in a way that could be argued, in Corderian terms, shows a lack of commodiousness:

> Rhetoric is love, and it must speak a commodious language,
> creating a world full of space and time that will hold our
> diversities. ("Argument" 428)

> To demand that others accept our particular experience is
> arrogance and dogma. To offer our experience to another may
> be the only plenitude we have. ("Academic" 317)

> Perhaps I am groping because I want to learn how this miracle
> [power of others' writing to take us into new worlds] is done, so
> that when *I* speak, others will listen, and not just because I'm
> responding to a need or a thought that they already have.
> Perhaps I want to make a miracle, though I surely won't.
> ("Hunting" 313)

Not only does Corder push toward a broad register—working both ends against the middle—but he is also transparent about the development of his points and arguments. He is well aware of the effects of his own most pithy phrases or turns of thought, so he imports sentences, tropes, whole paragraphs from one text to the next. I suggest this technique comes from artfulness, not from fogginess. In "Asking for a Text and Trying to Learn It," the essay where he explains how he came to write his own assignments with his students, Corder explains his process of keeping up with this new writerly load: "My common practice has been to scrape an essay together by the date the assignment is due. . . . I have taken parts of earlier papers, made more of them, and turned them in as new. I have rewritten rough drafts that I had submitted as finished essays in former semesters" (93-94). That he does the same—in effect, for effect, to good effect—in his published academic work speaks to me of a comfortable integrity: He is a writer working as he knows writers work, progressing by increment, not trying always to make new, keeping the bathwater, sometimes keeping the baby. To hell with you if you don't like or sanction it; welcome into his world if you do. As is normal in a writing life, his prose grows over the years—like a raindrop turning to hail—the attentive reader can hear what is imperative for him as he works up his strongest beliefs in greater detail, with more backing and passion. He's willing to say something again, to let his premises grow.

As I've mentioned before, part of Jim Corder's stylistic fingerprint—basic to his toolbox—is the inclusion of narrative, of story, often autobiographical in nature. This is not unexpected for someone whose published books mostly focus on history, biography, memoir. However, I think this is also intentionally part of his strategy for arguing that personal and academic writing are allies, not enemies:

> Once, when I was a boy—perhaps I was eight—I asked my mother what color my hair was. I can't remember why I asked, but I did. She raised up from what she was doing—washing clothes with a scrub board in a tin tub—scrutinized me carefully, and after a while said, "Oh, it's kind of a turd-muckledy-dun." The term signifies a muckle color that streaks off into dun color part of the time, into turd color part of the time. It was her way of reminding me that I shouldn't be occupied with my own significance. ("Notes" 102)

Not unaware of the power of memory and the way fact for one family member is fiction for another, Corder uses his family life selectively to make points. He is nostalgic for what is lost, willing to remember and praise it, but not really adverse to going on. He is elegiac, a memorialist, and he begs us to incorporate appreciation for the old in our considerations of the new. He is not opposed to new, but he plays it across his consciousness carefully, slowly. He regularly transgresses the scholarly/personal division, including story in essay: to the point and for a point. Not surprising, really, considering his arguments and interests outlined above.

Not surprising either since Corder's essays often center around existential considerations of self, asking: Who am I? Why am I alive? How will others know it? How can I (my writing, my rhetoric) communicate from one individual to another, no matter how well or poorly aligned we are in groups? Always, Corder argues, we nevertheless have to span the gap from one physical entity to another. From my thought to your thought and back again, recognizing all the while, the difficulty of that enterprise:

> Rhetoric, then, won't tell us all we need to know, but we can learn from rhetoric where to look—at the ways character emerges in language, at the ways worlds are constituted in individual discourses. Since we do speak different tongues and make different worlds, we face intriguing questions: Whom will we listen to? Who will listen to us? What transpires between us? How do we sing the Lord's song in a strange land? ("Rhetoric" 17)

In both substance and style, he works toward such spannings with the eagerness of one who wants "to leave tracks" but is also aware that to do this, it doesn't do to be overly concerned with one's own "significance." His sense of examining and savoring life is evident in the largess of his sentence style. Even as he has fun mixing high and low, he often does so in rhetorically intense—and effective—sentences.

The endnotes I have supplied for many of the quotations from Jim Corder's texts illustrate the complexity of his sentence style. When I first read his work, I found his prose style memorable, but I didn't know why. Looking at them with charts of schemes and tropes in hand, I discovered how regularly he preaches rhetoric through his textual practices, using onomatopoeic sentence structure (if there could be said to be such a thing) to generate and illustrate his thoughts. Most writers do; we strive to turn a phrase, but again, there is a sense—after reading the body of his work—that this deployment was intentional, part of the greater argument of Corder's systematic self-construction. Anaphora, antithesis, auxesis, brachylogia, diacope, direct address, epanados, epanorthosis, epanelepsis, hyperbole, inclusio, irony, isocolon, metapor, paradox, repetitio, all of the above and more. Corder's prose illustrates his intense engagement with language; he plays it, it plays him. These are the insights he shared with his readers and the opportunities he shared with his writing students, seeing them coequals in the great rhetorical game of being.

For me, the lessons of Jim Corder's writing boil down to these: that we should follow his lead in our own best manner; that we should write with our students; that we should work to explore and understand their rhetorics and our own; that we should write what we care about by identifying issues that are crucial to us as individuals, bringing them as gifts of plentitude to the tribe because each of our voices matters, because we are all in language separately and together. Writing teachers should be writing and publishing because it matters that we do, not how much we do. Jim's oeuvre illustrates this: His essays add up to an ethos enacted, a grand unified argument that rhetoric is as crucial as life, is life. That we are all rhetoricians. That we can learn to be more human through the practice of reflection and writing and rhetorical analysis. Jim Corder is radically pro the-life-of-the-mind. Keeping this in mind, we too should join in, writing ourselves from sentence to subject and back again. We should leave tracks.

NOTES

[1]Having written an appreciation of another compositionist whose writing influenced me and having been written to by one of that teacher's students who wanted to set me straight that the person I was praising for his ideas was not a person the student had enjoyed as a teacher, I'm aiming to be careful here and admit that I'm busy celebrating Jim Corder as a writer in my field. But I'd also hazard—on the basis of the ethos that he so carefully develops in his essays—that I would have greatly enjoyed

sitting down with him to discuss his ideas over coffee, thinking that we've all, no doubt, been a fine and a foul teacher to some student at some time.

²Anaphora, repetition of beginning words, and echoes of epanelepsis, ending a sentence or clause with the same word or phrase with which it began, or inclusio.

³Rhetorical question and two runs of anaphora: "I think" and "There is."

⁴Direct address.

⁵Antithesis, repetitio, "still here, still here, still here, still here," with diacope, words between the repetitions. Metaphor—turning (like a leaf on a tree: think of "that time of year thou mayst in me behold") and "leave tracks"; verging on hyperbole and cliché, "drifting into oblivion" and irony/paradox, "potentially noble, permanently errant."

⁶Isocolon, similar number of words or syllables used in the parallel repetition; refrain: "I'll be alone, and I'll be held accountable."

⁷Epanados, repetition in the opposite order, "with whatever purity they can attain/with whatever impurity they must reveal."

⁸Asyndeton, omission of a conjunction between words or phrases, "it is already normality, already truth, already the way to see existence" and anaphora "already, already, already"; ellipsis, "all discourse is ideological, we probably exclude our own [discourse]," also epanorthosis, addition by correction; irony, "When we remark, as we have become accustomed to remark."

⁹Auxesis, arrangement in ascending importance.

¹⁰Periphrasis, isocolon, antithesis, hyperbole, metaphor, personi-fication, synecdoche. . . ?; this passage—as is so often the case—is rich with schemes.

WORKS CITED

Corder, Jim W. "Academic Jargon and Soul-Searching Drivel." *Rhetoric Review* 9 (Spring 1991): 314-26.

—. "Argument as Emergence, Rhetoric as Love." *Professing the New Rhetorics: A Sourcebook.* Ed. Theresa Enos and Stuart C. Brown. Englewood Cliffs, NJ: Prentice, 1994. 413-28.

—. "Asking for a Student Text and Trying to Learn It." *Encountering Student Texts.* Ed. Bruce Lawson, Susan Sterr Ryan, and W. Ross Winterowd. Urbana, IL: NCTE, 1989. 89-97.

—. "At Last Report I Was Still Here." *The Subject Is Writing.* Ed. Wendy Bishop. Portsmouth, NH: Boynton/Cook Heinemann, 1993. 261-65.

—. "From Rhetoric into Other Studies." *Defining the New Rhetorics.* Ed. Theresa Enos and Stuart C. Brown. Newbury Park, CA: Sage, 1993. 95-105.

—. "Hunting for *Ethos* Where They Say It Can't Be Found." *Rhetoric Review* 7 (Spring 1989): 299-316.

—. "Learning the Text: Little Notes About Interpretation, Harold Bloom, the *Topoi*, and the *Oratio*." *College English* 48 (March 1986): 243-48.
—. "Notes on a Rhetoric of Regret." *Composition Studies/Freshman English News* 23 (Spring 1995): 94-105.
—. "Rhetoric and Literary Study: Some Lines of Inquiry." *College Composition and Communication* 32 (Feb. 1981): 13-20.

6

A Writer's Haunting Presence

Pat C. Hoy II
New York University

*I don't want to be constructed or interpreted. I want to
be known and acknowledged. I want not to be invisible.*
—Jim W. Corder (*Chronicle of a Small Town*)

&This morning as I sit in my New York City apartment and watch the
snow falling past my windows, I feel a deep yearning in my bones for the
hot pavement of summer streets in a little south Arkansas town that will
always be home. I'm wandering around the apartment trying to enter a
space where I can confront a ghost and I'm having a hard time of it. As I
sit, finally, to read a few lines from *Yonder*, just to get the feel of Jim
Corder's language and the rhythm of his sentences playing through me, a
deeper chill runs down my spine, and I realize that the socks I've pulled
on to ward off the shivers will not do the job. I realize too why I have put
it off so long, this act of recovery. I haven't wanted to face the loss.

A little more than a decade ago, after I had read *Hunting Lieutenant
Chadbourne* for Malcolm Call, the editor at Georgia Press, and had
completed my review, I reached immediately and unhesitatingly for the
telephone. I wanted to celebrate my reading of the book by getting myself
connected to the writer himself, a man I had never met in person. As I
read my review to him, Jim's breathy laughter crept into the spaces of my
reading, and when I had finished, we cried what I like to believe were
tears of joy.

When first I met the man face-to-face the following year at the
Conference on College Composition and Communication (CCCC), just
before we were to give our papers, the person I saw and heard was still
the man I had imagined while reading his book. But when we sat down
for drinks after our session, I found myself with a different man. I too
was someone else. Jim Corder and I might as well have never met, save
for one thing. Behind our silences, and our sharing of space, and the
dropping in of friends as we sat quietly drinking in a public space just
outside the conference rooms, there was a deeper knowledge of
something that we held in common, something that couldn't and
wouldn't be captured or repeated in a conversation, ever. What we knew

and what we had to say to one another that mattered would be there in the writing, forever. We both knew that then—when we weren't thinking about the future. Not then, not at that moment.

What I have to do now is account for the Jim Corder I have found in the texts—*Chronicle of a Small Town, Yonder*, and *Hunting Lieutenant Chadbourne*—the Jim Corder who, despite fashionable intellectual claims, will always be there in those last three books where, I believe, he tried to tell us most of what he had figured out about himself and his life's work. Virginia Woolf gives us Jacob (and his room), just as she gives us Percival in *The Waves* by hinting at their very nature. We barely get to see them, and yet, in the end, we are struck stone still by their death because we have come to know them somehow. Like those two men, Jim is there too, fashioned by himself in those books where he made himself known.

The seriousness of those playful books is highlighted by the scent of death: The methodology is experimental, but the stakes are high. This is not a Jim strapped by the formalities of rhetorical theory—not a Jim writing in the language of a profession gone berserk on divisive classifications and inflated language. And yet the Jim we find in these books is a serious, moody teacher, wedded to the classroom and the life of the mind as surely as he is wedded to the land and his peculiar version of the language of West Texas. There in the language of his everyday, we find his essence.

In these three books, there is much showing to complement the telling; more grace and elegance than rigor ordinarily allows; a knowledgeable and discerning writing *I*—standing somewhat aside while commenting on the other *I*'s (versions of himself at particular places and times in his life)—a writing *I* who reveals and explains what the stories themselves can only hint at; and, finally, a play of mind and spirit that could not adequately express itself in our professional journals where we so rarely catch glimpses of a man or woman's soul.

What this Jim is doing in these three books was so important to him that he could not feign objectivity; disappearing from the discourse was not permissible. Here in these books, he expresses himself, comes to rest . . . leaves himself for posterity. So doing, he bequeaths a legacy that calls us all to task. I want to delineate that legacy as best I can because I do not believe we can afford to lose him or to turn away from what he's trying to teach us, still.

The other Jim, the walking around one, is gone, except in our memory, and the writing Jim would be the first to say to us, "Never mind, memory is almost always wrong; it almost always misleads us about the truth we're trying to remember." That's what I think he would say. So I am not trying to write about the Jim Corder that everyone claimed to know, the Jim Corder walking around outside his texts: the teacher at Texas Christian University, the colleague, the father and husband and lover. I am writing instead about the Jim Corder in the books—his experiment and his vision, the things he thought about and what he made of them, whatever truth he discerned. Those of you who knew well the other Jim Corder, or claimed to know him, can decide what the two men had in common. I do not know, will never know.

The discerning, writing I, the I who stands apart and tries to make sense of things, no matter what the odds against him might be, is what Jim was interested in. That *I constitutes his legacy*, if we can only find him.

Chronicle of a Small Town (illustrated by Jim) looks, on the surface, as if it might be nothing more than a reminiscence, a partial history of a particular time in a little West Texas town called Jayton, the town Jim Corder came from. And so, perhaps as a matter of pure curiosity, he decided to go back to that town to see what he could learn about its history. But there is more to it than history because he was hunting something that would help him understand questions that were haunting him about life, about the nature of things, about himself, about writing.

The conceit for the book is simple enough. His evidence for the historical investigation would be the *Jayton Chronicle*, the local newspaper, and he would read those newspapers (all of them he could find) to check his memory about events, to try to better understand who he was as a young boy, or as an older man. And that is precisely what he does even though he cannot find all of the issues. When he sat down to work, he had only 241 of 676 of them, running from January 5, 1934 to December 12, 1946. Two whole years were missing: 1940 and 1941. But as it turns out, the gaps matter very little, given what Jim is actually trying to figure out.

Even the book as a representative history of the period works well without all of the issues. We might just be reading the story of any small town in America on the one hand, but on the other we learn the intimate and particular details of Jayton, in that corner of West Texas, in that time: the weather, the idiosyncratic life of the townsfolk, the location and disappearances of the stores, the rhythm of life, the slant of the light on the hills, the ups and downs of a farmer's life, the centrality of sports, the movies they saw, the war and the airplanes and the deaths. But this is not just an ordinary history book, not, as the title seems to suggest, just a chronicle. The book itself is an investigation into the limits of knowing, just as it is an experimental essay, a book-length familiar essay, that probes beneath the surface of the events it recounts to reach a deeper understanding about writing and knowing and living.

Jim Corder challenges us from within this book as he meanders his way through the newspapers and the events they record. Researching, he wants us to know, has a bit of wandering in it. It is neither directed toward certainty nor destined to uncover all the truth—not because there is no truth to uncover but because the final truth is so hard to find. He challenges us to think deeper and more expansively about what research really means and about how we inevitably become a part of the research that we conduct.

The researcher in Jim's scheme of things is trying "to get into as many situations for seeing as possible." But there is no clean objective way to conduct our work, no way to separate ourselves and our interests from the investigation, and yet there is a value in what we do, in what we find. Our opinions and memories are "poor substitutes . . . for whatever the truth may be, but they're mostly what we have, tentatively or otherwise, as truth, and can rise not just from thoughtless perpetuation of prejudice and faulty memory, but also from such study and thought as we can manage" (*Chronicle* 19). Such study and thought constitute our research, and that research is far more complicated than most definitions acknowledge:

> Research, too, is opinion and interpretation at last. No research methodology is divine; none is the sole route to truth. A research methodology is a human rhetoric that lets us create structures of meaning. Such rhetorics are not in the nature of things; they are the spoken, written, created nature of things, deriving from the utterances of a community, much as facts become factual through the agreement of an assenting community. Knowledge is not a sacred finding, forever fixed, but the product, temporal and temporary, of human insight and judgment. (19-20)

"Such rhetorics are not in the nature of things; they are [themselves] the . . . created nature of things." We are the creators. We render the judgments; we reach consensus among ourselves; we make the facts factual; we re-see and revise; we keep looking for the truth that is so hard to find. And there is no single path to that truth; no sacred methodology that will take us to it; no disinterested state of mind that will yield it up. In fact, our interest, our concern, may lead us closer to it. Jim himself shows us how.

As we watch this fascinating researcher check the facts of his own life, as we watch him face the surprises that research delivers from day to day, as we listen to the stories he tells about these surprises, we begin to see how his interest in the material and in his own life prompts the investigation and eventually leads to corrections in memory, just as the process raises, at the very same time, new questions—continual discovery, continuous revision . . . a tentative grasp of some long-buried truth accompanied, almost always, by the blurring of another certainty. Bound in the narrative of his own experiences, the researcher finds himself simultaneously enlightened and stymied by the discoveries and surprises so inherent in the work itself—in life itself.

In this chronicle Jim is at the very center of his investigation, revealing in great detail all of the truth he can muster about these discoveries and questions, giving us both the joy and the frustration that accompanies the search for knowledge—knowledge, what Cardinal Newman called an "acquired illumination."

One of the primary things that Jim wants us to know is how this search for knowledge always leaves us knowing that we need to know more than we can find at any given moment, more sometimes than we can ever find. This motif of difficulty becomes a refrain that pervades the book's design and carries over to *Yonder* where we see it in a different light. Listen to Jim, caught up in a moment of such difficulty:

> I cannot account for what happened next. Perhaps it was a simple sequence of associations. Perhaps it wasn't simple. When you go back in time looking for something like the truth, sooner or later you come to abut the place of which there is no knowledge, whether that is prior to birth, in inadequate records, in mistaken memory, or elsewhere. You come to abut that place in any search. (106)

Later in this book, stuck in another moment of difficulty, Jim comes back to these same thoughts. Always, there is for him that place of abutment: "When you look everywhere, look back or out or sideways or in or down or up, wherever that is, you come at last, in any direction, to the place of which there is no knowledge, of which there can be no knowledge, come to that place but not into it, come only to abut it" (*Yonder* 166-67). He despairs of not knowing how to look everywhere, of not being in the right place, of not finding what he wants to find; it becomes, finally, "a search for what is not, for the thing that never happened, for the place that never was" (167). And yet there is no giving up: "That doesn't mean there is no reason to look. I'm also there, and the others are, and the places are, and there are always reasons to look, to get into as many situations for seeing as possible" (167).

What is driving this inquiring mind is a deep sense of loss and a chronic nostalgia for what was, or might have been, coupled with an insatiable desire to know more, always to know and understand more. He speaks clearly about stepping into modernity—a move that is always accompanied by loss and uncertainty. But while there may be pain and loss in the movement, he is committed to it—"always shifting from the pastoral (even if it lasts only a moment) to the urban, always having an Industrial Revolution in every life, in every generation" (*Yonder* 168). He knows that this is an archetypal pattern, that

> we are always, if we notice, in exodus, *volkswandering*, always, if we notice, in an age of discovery, or upon the rim of discovery, always making, or trying to make that leap into modernity, always feeling the loss that the leap into modernity brings, always recovering, or trying to recover, always on the edge of Holocaust, or there, soul by soul, generation by generation. (169)

And so Jim struggles to remember "the leeks and the onions and the garlic" even as he contemplates, or gets caught up in, the great sweep of History, just as he willingly makes the leap that will keep him from settling into that "one place for looking" that will keep him out of the many places—the one place that will put him in blinders. Too much comfort leads to blindness and not knowing (*Yonder* 169).

If *Chronicle* gives us glimpses into that "exodus" from the muddled but comforting past *into* modernity, *Yonder* throws us into the maelstrom of a present that can rip a person apart. *Yonder* is first and foremost a book about the fall into History—our fall, as well as a particular man's fall into History at a given moment in time. The title and the subtitle could easily seduce us into thinking that we're about to enter a futuristic paradise out there, over yonder, somewhere—*Yonder: Life on the Far Side of Change*; instead, we find ourselves immersed in the turmoil of a man's personal life where we are made to experience the pain and suffering that accompany change.

We enter a period of Jim's life outside the text where change destroys tranquility, creates self-doubt, invites self-indulgence, pulls and tugs on the heartstrings. His story of divorce and, then, the story of a new relationship, along with the accompanying stories of conflict within his larger family, those stories brought into the text—and because of the way he tells them—leave me gasping for relief. Not because I am squeamish about the revelation of the personal (my favorite form of personal expression is the familiar essay) but because I do not choose to let my readers see me suffering. As he struggles to present the truth, I wince because I would not use the personal the way he uses it. I do not make my suffering the subject of my work. But I have to remind myself that this is a long, searching familiar essay, that Jim can use his experiential evidence any way he chooses. My job is to follow and understand, and when I look with him beyond the tales of suffering, I find a more important idea taking shape through those stories. I suspect that for him it is the central idea of his later work because it has everything to do with survival.

No familiar essayist worth his or her salt tells us personal things for their own sake. The personal is there to highlight and reveal the archetypal. We tell our stories to others because we want others to be able to claim our stories as their own. The particular always invokes the universal; it must, or the writing will not pull us beyond the mere facts of a single life. The narrative will be bound and restricted, interesting, if it is interesting, only for its own sake. Jim understands such limitations perfectly, knowing full well that he is working the personal against the backdrop of history even as he draws, simultaneously, from the great reservoir of archetypal stores:

> I wanted this book to be a scholarly sort of work written in a personal sort of way. As deliberately as I could, I made it personal, but I don't much think it's ever just personal. I wanted change to be known as immediate and personal; I wanted history to be present, to hurt. . . . I hoped I might see what there is in my circumstances that would let me lift the personal up to public scrutiny, bring the public in to private scrutiny. (*Yonder* x)

He had, of course, done the same thing in *Chronicle*, but there is a vast difference between the story of childhood memories gone awry and the story of family upheaval; the writer's focus makes all the difference. In *Chronicle* we toy with the correction of memory, the broader implications of personal discovery, and the limitations of knowing. In *Yonder* we peer directly into the heart of a man's darkness, for a long time, and we suffer with him so that we can reach a deeper understanding about survival.

We peer, if we can stay with it, into a revelation about lasting. What this book wants us to know, above all else, is the importance of life on the printed page. That life can last, but lasting in the text is almost as complicated as lasting outside the text. History and, at this moment, poststructuralism have made it so, but neither of these forces deters Jim; they simply make his task more difficult and more interesting.

Jim tells us straight out that "nothing exists until it exists on the page," and then he offers a corollary: "I have imagined, or wanted to imagine, or believed I imagined that nothing related to me will count unless I get it outside myself" (*Yonder* 75). And so this book, this particular book, is a deliberate act of confession, an act of proclaiming and designing a life in the text by selecting from a life outside the text. They are not the same, those two lives. The life in the text, crafted and confessional, points to that idea about "life on the printed page." To last, Jim Corder has to get that life outside himself and into the public domain and into the text; he has to do that to make his life count. But consequences of all sorts hang heavily over his individual acts of re-creation.

Let's begin classically, as Jim does, as a way of understanding the consequences. He puzzles over Harold Bloom's insistence that "there are no texts, but only interpretations." Jim actually tells us, "I shake fearing my own existence, for the narrative I've made also disappears" against such an idea. But countering Bloom, he insists on the existence of the texts themselves: "The text we make [through reading and interpreting] is always tied to, rooted in, that other text, gets its occasion from that other text. Texts exist: they unfold or keep secrets. They go on unfolding and enfolding while we're watching. They slip away" (*Yonder* 18-19). Yes, they slip away from the reader, but they do not slip away as texts. Meditating on Bloom's challenge, Jim shores himself up with a review of some of the central terms of classical rhetoric. The most essential and critical for him are *oratio, narratio, exordium, peroratio,* and *refutatio*.

We must hear Jim at length on these matters if we are to understand just what rhetoric has to do with survival. As we take up his argument, he has been thinking in the text about the simpler definitions of the classical terms. He turns here to his own amplification:

> But I was a long time learning that the *oratio* is not just a way of outlining. When I looked again, I learned that the form tells us something about ourselves, more than I had thought.

Looking, I thought I began to see that *narratio* is not just a part
of speech; it is also the form's testimony to what always
happens when we speak and to our attempt to lift the inevitable
up to the conscious. We're always situated somewhere when we
speak, whether or not we have placed ourselves deliberately.
Every utterance comes from somewhere, exists in a setting.
Every utterance carries a history. *Narratio* calls us into our
history, calls our history into our text, tells us to tilt our ears
toward the history carried in the other's speech. I slowly begin
to learn that *narratio* is not just part of a strategy but is also,
more significantly, our recognition, acknowledgment,
insistence that we create a world when we speak, staking out
claims, making a history. We are not representing reality but
making it, and *exordium and peroratio* want us to know that
we are makers, deliberately commencing a reality and closing
our account of it, not at *the* beginning and *the* ending, which
can't be found, but at made sites of convenience, where speaker
and hearer can convene.

I begin to learn a little about *refutatio*, too. We are not
required to be disputatious, but the text of centuries teaches us
that we do sometimes make our meaning against others, and
sometimes beside and toward and for others, and always in the
midst of others.

The *oratio* turns out to be a remarkable exemplification of
ourselves. It is our drama and our sermon; it is the human
dance. I didn't see that the first time I looked, or the second, or
the third, or many times after that. (*Yonder* 19-20)

Jim wants us to see that when we write, when we tell our stories,
whatever they may be, in whatever form, when we use them to create our
disputations, we are creating ourselves, making something of ourselves
within the history that we inevitably bring into the text. We are making
history as well. The text, its form, reveals us, but there is more: "[W]e are
making a world." Our *oratio* is "our drama and our sermon; it is the
human dance." We are known, or can be known, both by those words we
choose and the form we choose for their expression. And even in a
classical structure as tight as the one Jim uses for illustration, he reminds
us that there is room for variation, for self-expression. There always is.
There must be (or we are not dealing with the living word).

I wish that Jim had told us more about form. I want to know more
about what he meant when he claimed that the *narratio* is the "form's
testimony to what always happens when we speak and to our attempt to
lift the inevitable up to the conscious." He tells us clearly what he thinks
the form does with history, but I want to know more about the form
itself: its shape, its variability, its idiosyncratic way of expressing, forever,
the writer himself. I know that it has to do with storytelling, with the
peculiar and particular way that one person tells his story—creates his
refutation, exordium, . . . I know that the writer is making something of
himself or herself as he or she performs those acts of self-expression. But
I want to know more of what he knew about the form itself and the way it

both permits the expression and helps define and give substance to the self that is expressed.

There is a further hint elsewhere in this book about surviving the fall into the maelstrom that is history. We find Jim again in a classical frame of mind, fussing with the poststructuralists, even as he admires and acknowledges their thinking. He could embrace contraries without the slightest hesitation. He would call it living with "competing rhetorics." Let's hear him talking directly to us, about this important matter: "I had wanted to believe, you see, that ethos, the character of a speaker or writer, is real and dwells in the text left by the speaker . . . and that I could find it and perhaps even leave evidence of my own character in stray words thoughtlessly left behind" (*Yonder* 53). That is what he had wanted, what he wants, and yet he is troubled in the face of this news from the poststructuralists who have robbed him momentarily of that comfort of believing that he can last in the text. He makes clear again how high the stakes are, for him: "If ethos is not in the text, if the author is not autonomous, I'm afraid that I've lost my chance not just for survival hereafter (that happened some time ago) but also for identity now." He goes on to elaborate the consequences: "[L]anguage writes us, rather than the other way around, and interpretation prevails rather than authorship" (53-54). He despairs, knowing that in his own words "'they' want me—and you and all those others—to die into oblivion if we should manage to scribble something, never to be reborn in a voice for some reader, but to vanish before that reader's construction/decon-struction/reconstruction of the small things we leave behind" (54). Against such annihilation, Jim calls necessarily for a witness:

> Will anyone notice that I may be here, that this is the way I talk, that this is what in my mind passes for thinking, that this may be myself? Life is real, and the artificial compartments we create for it don't work. What gets said in one place keeps slopping over and meaning something in other places. I'm talking about my own identity now, the nuttiness that is mine. (*Yonder* 54)

He acknowledges that writers select the details that go into the texts they create, deciding, as Annie Dillard reminds us elsewhere, what to put in, what to leave out. Those writers can, and often do, conceal as well as reveal. But despite that fact about concealment, Jim is not ready to give up on being a "post-Gutenberg man" (*Yonder* 75). He is not ready to acknowledge disbelief in the importance of the person who has inscribed himself in print. The truth, "if there's any truth out there somewhere or in here, it's in the character of the small particulars" (75-76). He worries only that he has not paid enough attention over time to accumulate sufficient small particulars that will allow him to tell his story true . . . and that memory will fail him, as memory almost always does, without those particulars. He cannot give up on the text; it has everything to do with his own survival.

There is in *Yonder* yet another dimension to this inquiry into textual survival, but I want to save it for closing. It has to do with those "competing rhetorics." I can't tell that story yet. For now, I stick with memory because it is again memory that informs the final book: *Hunting Lieutenant Chadbourne*. Already, near the end of *Yonder* (177), Jim is being drawn toward Chadbourne—and to a book that will be a work of impeccable scholarship. In *Chadbourne* Jim perfects the artful marriage between the familiar and the academic. Having hunted himself in the two earlier books, he has learned how to conduct the search, and he has learned as well about the limitations of finding. In the process he has become a more graceful writer.

Lieutenant Theodore L. Chadbourne first caught Jim Corder's attention on an historical marker between Abilene and San Angelo, Texas. Chadbourne had been stationed in the vicinity of this marker as a young lieutenant and was subsequently killed at the battle of Resaca de la Palma, May 9, 1846, in the war with Mexico. A fort had been named for him. Something about that man—his youth, his premature death, his short life in West Texas—something about the marker, took hold of Jim Corder, and he decided that he had to see if he could find Chadbourne, dead almost 150 years. On one level this book is the narrative (the *narratio*) of that search; it is the intertwining story of Chadbourne's early demise and Jim's research, with all of its ups and downs and twists and turns as he both fails and succeeds in his effort to track down the missing man.

There is more. If in *Chronicle* there was a fascination with memory riding alongside Jim's reading of the *Jayton Chronicle*, and in *Yonder* Jim's obsession with change was underscored by his deep concerns about existence, there is in *Chadbourne* a relaxation of the struggle that in those earlier books kept us always conscious of the difference between the ostensible work of the books and the needs of Jim's psyche. In *Chadbourne* I detect no such separation, anywhere, and yet the two levels are there, flawlessly integrated.

The actual hunt for Chadbourne is complicated not only by the scarcity of the historical record but by the way memory necessarily affects the findings. Early in his search, Jim sees Chadbourne's sword belt in a museum showcase, but he remembers that he has also seen Chadbourne's tunic in that same case. He discovers later that he has not, and so he uses the occasion of this error in remembering to take us back to the *Jayton Chronicle* and to various other misremembered events in his life so that he can raise once again the vexing questions that he has been pondering about memory and imagination, raise them this time as a concern about the accuracy and value of the research that he is conducting on his historical subject.

His thinking now is more complex than it was in the earlier books, and he is more at home with the poststructuralists. But he has his own ideas. He still knows and seems to believe that "memory is the tale we tell of ourselves; it is the selves we keep constructing in our continuing narratives" (*Hunting* 10). But there is a new twist. He no longer sees the failure of his memory as a personal failure, nor does he see, I believe, the failure as a detriment to the history that he is constructing. Even with the poststructuralists on his mind, he is more settled about what he has figured out. He wants us to know that whatever is given as truth in fiction or memoir has its own integrity: "Whatever is given doesn't have to measure itself—nor do we have to measure it—against what might have happened somewhere, sometime. If neither author nor reader but text is our focus, then textuality is the issue and there need be no coincidence between whatever is given as text and whatever might be another reality" (11). The text itself is sufficient. "Memory is already sanctified and sufficient, even when it very plainly is not" (11). As he wonders about whether we can ever "get situated to see, or resee, what may have happened, who may have been, where we might have lived," he continues to find ways to value that which we construct. Memory serves us ill only when "we expect it to correspond exactly to a reality out there" (18). Somehow that work that we do as writers, using memory and imagination together, leaves us with a residue that closes in on truth. The work, despite faulty memory and the imposition of the personal error, can lead us to Chadbourne.

What this tentative confidence seems to ride on is the fact of the subject's reality. As he thinks about his troubling colleagues in the English department, Jim thinks finally of the fact that they are real in some sense that is independent of his mind's thoughts about them. They are real the way a text is real, the way Lieutenant Chadbourne was real 150 years earlier, independent of anyone's perception of him. And so what Jim wants to know about the subjects he studies is "how to read them," how to "go toward" them. He reminds us, "I can't see or think or live outside my perceptions, but all those others out there are not just ghosts inhabiting my perceptions, though they are that, too" (*Hunting* 97). And yet what is important for the writer and for Chadbourne's survival and Jim Corder's survival and mine and yours is that the writer's perceptions about them count. In the case of Chadbourne, Jim reminds us that except for "sparse records and a double handful of letters, he doesn't exist except in my perceptions." Jim has no choice but to "go out toward the others, take their testimony first, hear his" (97).

That is what Jim does; he goes out toward the others: He takes what is left to posterity by Miss Susan Miles, the local historian who had already fallen under the spell of Chadbourne's memory; he goes to the written records about forts and their naming; he searches, through various intermediaries, the government records in Washington; he goes to West Point to consult Chadbourne's cadet records; he corresponds with and swaps evidence with descendants; he researches the lives of Chadbourne's West Point classmates, those who were in Texas with him and some who were not; and he reads, finally, the primary sources,

Chadbourne's own letters, twenty-two of them, left by Miss Miles in the box containing the fruits of her own research.

We know just where Jim puts his trust in this long trail of research from the way he orders his own narrative about it. If Chadbourne is out there anywhere, apart from Jim's perceptions, yet bound up in them, he is going to be, according to Jim's reckoning, there in his letters. So we do not get to see Jim read those letters until near the end of his own narrative, even though he had the letters almost from the very beginning. He waits, not to be disingenuous or rhetorically savvy, but to let us see as he saw.

Near the beginning of Jim's story of the letters, he tells us what we could easily anticipate, having come to know him by now through his own work: "History is not out there; it's in here, where I am. I wanted to save the letters for last because I wanted to believe that he was in them, that he was there" (*Hunting* 146-47). Whether Jim found in those letters only those aspects of Link Chadbourne's personality that matched his own I will never know, but whether I am seeing a reliable rendition of Chadbourne, I have no doubt. The evidence gives me confidence in Jim's reading of it, and when I come to the end of this magnificent book I know that ethos—"the character of the speaker or writer"—informs my understanding of Chadbourne and all that Jim Corder has to say about him.

I expect Chadbourne's "good humor" along with the "consistent fairness in his way of reporting." It comes as no surprise that "he puts personal limits around what he tells us—it's always *his* perspective; not universal truth, not dogma, *just what he has experienced so far*" (*Hunting* 151). And when in Jim's favorite of Chadbourne's letters—where the young lieutenant writes to the adjutant general of the army about being passed over for a promotion—where Jim praises Chadbourne for the "elegance of his charge" and for the "courteous, insistent force in the letter," I know that I have found the two men on a common piece of ground (154). I am not a bit surprised that in other letters Chadbourne is "solicitous of his father" or shows "a kind of directness, an unflinching honesty in the way he faced life," not surprised that Jim found no "trace of deceit or rancor" in the young man's words (155, 157). I am not surprised because I have learned in the books that those are traits that Jim Corder values. They say as much about Jim as they do about Chadbourne. But Jim does not make them up. He does not ascribe them to Chadbourne without the evidence. We know about them because Jim has the capacity to see them, to point them out to us. We know about them because his perceptions permit him to recognize and honor them.

Knowing these things about Chadbourne, we can imagine why Jim stayed with him, why he wanted to understand his premature death, why he took the time to conduct the research and write the book about him. He wanted, finally, to make him last. He wanted us to know the young man as he came to know him, and he wanted us to understand why such searches and such writing are necessary.

I have, of course, been following his example in this chapter, trying my best to find the man in the texts. Like him, I trust the texts and understand the importance of trying to give you a clear account of my perception of those texts without having to worry about whether I've gotten it all right. I can't get it all right, but you, Jim, have shown me that I need not. I need only get my perception as clear as I can get it and give the others a chance to see my evidence, to second guess me, to pit their rhetoric against mine in this world of "competing rhetorics" (*Yonder* 164). I have been with you long enough during the reading and writing about these three books to begin to hear your voice at the back of, or next to, mine. I hear you warning me: "Your construction of me is myself against your past, but not myself" (177). And I know what you mean, but you are there just the same, as the text is there, and whatever part of my past I've brought into these reconstructions and judgments is a part of me that you have called forth. Without your help I would not be thinking of you. I too am nostalgic in the face of change and loss, and without your help I could not begin to locate you. I have found in you what I like in myself just as I have found some differences, but my effort to hunt you down has not been an effort to distinguish us one from the other but an effort to understand you and pass you on to others as best I can. I have tried to do what you say we must do, and that is to take the time to notice the details, the small particulars, and to get them down in a language that obscures neither of us.

I remember what you've told me: "If we're to know each other from rhetoric to rhetoric, we're probably going to have to risk showing each other our lives; not pronouncing toward each other, but revealing whole structures of meaning that tell us how we worked our way out of and through our past, choosing and not choosing, making a shape and style that give residence to meaning" (*Yonder* 172-73). What we need to do is try "to live in our own voices" and, as you have advised us, "inhabit them not as owners but as sharecroppers" (173).

I must confess that I wish the other Jim, the one outside the text, were here now for a brief visit because I would like for him to know how I have looked for and claimed the essence of the Jim who will last, who will forever be here in the texts, an indestructible part of them. I yearn for outside Jim's throaty laughter and his willingness to sit with me in relative silence as we honor our other, more transient, selves.

WORKS CITED

Corder, Jim W. *Chronicle of a Small Town*. College Station: Texas A&M UP, 1989.

—. *Hunting Lieutenant Chadbourne*. Athens: U of Georgia P, 1993.

—. *Yonder: Life on the Far Side of Change*. Athens: U of Georgia P, 1992.

7

Finding Jim's Voice: A Problem in Ethos and Personal Identity

George E. Yoos
Professor Emeritus, St. Cloud University

ᐁWhen I read Corder's *Yonder*, trying to place it, either as a literary or a rhetorical genre, as Corder expresses it, "to give it a name," Thoreau's *Walden* easily came to mind. There is something anomalous about naming a category of writing that fits *Yonder* and *Walden*. Are they memoirs? Or are they just examples of personal journals? Are they celebrations of geography and place? Are they simply compilations of somewhat thematic essays? Or again, are they just philosophical meditations in the spirit of Montaigne? What is it about both texts that make us want to think of them as literary works? I suggest that we can regard them as being all these things, and more, especially in having literary aims and purposes.

I find that there are four kinds of criticism going on in scholarship within English departments. Historical criticism does biographical work about authors, their social context, and times. Literary criticism seeks to point out and establish the merit of literary authors and their works. Rhetorical criticism seeks to evaluate the effectiveness and even the merit of the persuasive agenda of texts. Cultural criticism attempts to look at writing from multicultural perspectives, including the cultural and ideological context of authors and critics. It is interesting to engage *Yonder* from each of these four critical standpoints—historical, literary, rhetorical, and cultural.

Both Jim Corder's and Jim Kinneavy's patterns of influence in Texas on rhetoric and composition drew my attention first as editor as we expanded from a newsletter into the *Rhetoric Society Quarterly*. What struck me early about their work was that Corder appeared to me to be the much better writer of the two. What was my basis for this judgment? What struck me was that my judgment was incongruent with many of the things that made me early in my academic career want to leave English studies and go into philosophy in graduate school. Good writing for me had to be not just stylistically good, it had to be substantive.

Early as a high-school student and again in junior college, I was mystified that English teachers graded student writing solely with handbook writing standards. I was under the illusion when young that to write well was to have something substantive to say, something intelligent and thoughtful to talk about. That you said something well was secondary to having something to say worth saying.

As a teacher of philosophy, I was often struck how students could write cognitively empty philosophical papers with good writing skills. What the students were saying had no depth, no attention to presumption, no careful analysis, little train of inference, and little assessment of facts or careful observation. Sonorous and flowing sentences about political and religious convictions may be blessed with conviction, but to anyone questioning those convictions, testimonies of faith and conviction have little to offer in the way of reason or explanation. Philosophy seeks warranted assertability.

Today I see the issue framed somewhat differently. E. D. Hirsch in *Philosophy of Composition,* in using a standard of relative readability, relegated invention in writing to saying things with relative clarity and simplicity. He separated the standards of good writing from the invention of substantive issues. I do not believe, contrary to Hirsch, that you can easily separate invention that ends up with a readable style from invention that generates substantive and persuasive content. Stylistic invention and substantive invention complement each other. And I am inclined as a philosopher to give the prize to substantive invention in developing writing skills. Why was I not looking at Kinneavy's writing from a substantive and cognitive point of view, which seemed to be his forte? Was I not giving priority to style in judging Corder the better writer of the two? Was I not confusing literary merit and cognitive merit in giving my nod to Corder?

And then it occurred to me that Corder's effective management of voice was what attracted me to his writing style in the first place. Finding voice for him was central to rhetorical strategies in the teaching of student writing. It enabled him to better bond with students. And Corder's focus on voice, ethos, and persona, I want to contend, has the net effect of introducing tacitly literary standards and not cognitive standards into our perception of good writing.

Corder, sometime in the 1970s, asked me to work with him, to do something on ethos and ethical appeal. At the time I recall sending him a number of pages of speculative explorations of the topics. Strangely he did not respond. At the time I was a rank amateur in rhetoric. I was all too willing to work with someone in rhetoric who had an established reputation. However, I was somewhat taken aback when he did not respond to the materials that I had sent him. And Corder never ever again brought up the subject of my positive response to his overture in any of our subsequent correspondence and friendly encounters.

My surmise at the time was that he found me, as an analytical philosopher, methodologically incompatible as a possible working partner. After reading *Yonder,* I think that my suspicion was correct. We were intellectually miles apart. But I have in no way disagreed with Corder's persistent assessment of the importance of ethos and ethical appeal as rhetorical strategies. Both, I believe, are dominant in sustaining reader interest and attention. But I would want to qualify his overall assessment by saying that the importance of ethical appeal depends upon what kind of writing we are talking about. Corder never seemed to be talking about some of the practical and functional modes of discourse that Kinneavy discussed.

Lately in exploring the characteristics of memoirs and examining the rhetorical uses of narrative, I have concluded that there is something inherently literary about memoirs as a rhetorical genre. First-person accounts in memoirs generate powerful personae. Storytelling overwhelms reasoned argumentative discourse by the richness of its rhetorical indirection and contextual implications. The author-narrator role in memoirs, as in lyrical poetry, tends toward fictive posturing. And voice and narrative in memoirs take on the expressive and dramatic features lacking in straight expository prose.

Recently, memoirs, although nominally nonfiction, have become highly competitive with fictive storytelling. A problem with memoirs, and unlike with biography, is that they are like historical fiction, not truly history. Author-narrators only present evidence for historians or their biographers. They do not write history. And the authenticity of what the author-narrator says in a memoir about self is always an open historical question.

Corder thus for me has been one of those rhetoricians, such as Wayne Booth, where voice and ethical appeal have been part and parcel of their views on rhetoric and rhetorical strategies. I have asked myself, time and again, how crafted and authentic is the voice of Booth in his writing? Surely as a careful writer, he has rigorously crafted his own persona in his writing. And how authentic is the voice of Corder in his writing? How carefully has he fashioned his own persona, especially in *Yonder*?

Noteworthily, Corder does not want to think of *Yonder* as a memoir. Yet in some ways it is one. But it fails as a serious memoir as Corder refuses to disclose important events in his personal narrative that would contribute to an understanding of what he was all about in his quest for self-awareness and personal identity:

> Do I fail to show enough, to tell enough? Do I show only effect, not cause? If I don't provide enough referents, will you wonder why I act and think as I do? Have I not made plain what I react to or against? I haven't explained a divorce. I haven't explained about the psychiatric ward in Galveston. I haven't explained why I am lost to my daughter, and she to me. What is hidden among what is revealed? Do you need to know more? I may not tell. (*Yonder* 56)

That things are revealed in what is hidden in *Yonder* raises questions about the authenticity of Corder's narrative. *Yonder* reveals his struggle in trying to come to terms with himself. And such an intense effort at self-discovery well illustrates Corder's problems in thinking about ethos, ethics, and ethical appeal. Given my own preoccupations with using ethos *as* a mode of engaging and negotiating, I am personally apprehensive about my present deconstructive commentary about Corder and *Yonder*. I too am wrestling with my own authenticity in saying what I have to say about Corder.

I am fully aware of the ethical issues in any personal criticism of *Yonder* in a collection of essays intended to laud Corder. And that is why I have spent so much time reading and rereading the book with loving care. There is a point in time, I believe, where interpretive criticism reaches the level of an author's own editing and revision. I do not know if that has been the case with my own reading of *Yonder*, but I think that I am close to the bone in dissecting what Corder had to say. I am absolutely intrigued with *Yonder* and its integrity as a work of literature. But I have critical reservations about *Yonder* as serious critical discourse advancing rhetorical and cultural views. With that said I especially want to take note of Corder's following cautionary warning against any one who is about to criticize him:

> If I criticize a single action or utterance of the other, that is the other's identity I speak against, not a separable item that may be regarded coolly, but an item that connects with, reverberates through, all of the other, hence *is* the other. If the other criticizes a single action or utterance of mine, that is my identity that the other speaks against, not a separable item that may be regarded coolly, but an item that connects with, reverberates through, all of me, hence *is* me. ("When" 53)

Corder begins his narrative in *Yonder* describing events occurring in Washington in June 1988. He begins by revealing his personal crisis with his oldest daughter, Cathy. And he immediately follows that first narrative with a description of his state of mind while once again drinking at a bar in search of the perfect peanut. *Yonder* begins by introducing us to his personal problems. Oddly, it concludes with a narrative occurring more than a year earlier in the spring of 1987 that has been somewhat compromised by narrated events that occur after the event. The conclusion of *Yonder* is about a trip to San Angelo to tell Mindy, his second daughter, and her family about his plans to remarry. He expresses in conclusion a glimmer of hope for the changes to be anticipated in remarriage. But this hope does not resonate with accounts in previous narratives about events subsequent to the San Angelo trip. Narratives are out of joint.

Throughout *Yonder* there are hints of alienations with parents, brothers, and sisters. There are obscured narratives about his relationships to his first wife, daughters, and son. Something dramatic has happened with his daughter Cathy. We are not informed about what led to the divorce or about the date of his remarriage. Such data are important for a curious reader interpreting and sequencing the events of Corder's life. They are surely important if we want to understand his motives for writing *Yonder*. Why the lacunae between large transformative events? Why the evasions? How strange are the inconsistencies in his states of mind after the events in the San Angelo! What is before? What after? He hides the order of cause and effect. Is he using narrative ambiguity for literary effect? Or has he chosen not to reveal too much to protect the living? If so, hasn't he said too much already?

Yonder ends for me much in the way as does *Sons and Lovers* by D. H. Lawrence. A protagonist walks down the street with none of his problems solved. We have the same sort of closure in author-narrator Robert Pirsig in *Zen and the Art of Motorcycle Maintenance*, which like *Yonder* is also a reflective narrative-memoir about personal problems with a great deal of background reflection on the subject of rhetoric. Likewise as in *Yonder,* there is no closure. *Yonder* ends in nostalgic, reminiscing lament. He is still drinking. He is not looking forward to any great future. He is searching for what is vanishing, which includes himself.

In *Yonder,* according to Corder, each of us has our own time and place, our own incompatible languages and rhetorics. There can be no ultimate mutual understanding between us. Languages and rhetorics make it entirely impossible for us to understand fully the other's meaning. Each has very little hope that he or she can transcend our personal and cultural limitations in communicating. It is as if he were saying, "Try to understand me, but you can't." Yet Corder still tries.

All discourse for Corder in *Yonder* is fundamentally synecdoche, only fragments of referential communication used to communicate meanings and understandings. Discourse by nature is selective and limited. And in *Yonder* Corder displays only bits and pieces of himself. Given these fragmentary synecdochic limitations to our understanding of what Corder says about himself, how are we then to treat rhetorically his selective use of narrative fragments, especially when we know that in all probability he is hiding or denying things? We can be fairly confident from internal evidence in his text that he suffered from severe and chronic depression and that there is substantial evidence that he was an alcoholic. Moreover, the organization of *Yonder* reflects Corder's expressed confusion about his disordered conception of events and time. How purposeful were his arranged inversions of time in his nostalgic narratives? Are we dealing with poetic license?

Many dates and descriptions of events given by Corder, if accurate, would provide an historical critic with a detailed map of what probably happened in his life. Corder leaves a long trail of textual evidence for us to make reasonable suppositions, suppositions that question what he says about himself, and whether or not he was in good faith, either with himself or with his readers. Note the following narration of events and what any competent psychotherapist would most likely make of his saying such things. The events follow the departure of Roberta on the previous day:

On Sunday morning, October 11, 1987 at about 8:40 Patsy
called from the airport to tell me Cathy was gone, would reach
Washington around 2:00 P.M.

Don't leave me in the dark, Cathy. But I left you.

That Sunday evening it was uncommonly quiet
everywhere—on campus, where I had worked most of the day,
at Abernathy's where I went to drink. For a while I was the only
customer. Few things stirred.

Cathy is gone. My family is gone. I am gone.

By Monday evening, October 12, 1987—have I got the time
straight?—I'm in reasonably bad shape and isolated from the
world. All is quiet. The world is elsewhere and I'm in jeopardy,
and I think how much nicer it is to be drunk than it would be to
be an alcoholic.

Have I got the time straight? (*Yonder* 35)

Note he does not reveal in the text the identity of Roberta and Patsy.
He does not reveal to the reader that his first wife was named Patsy
except by implication. Nor does he indicate the time of the divorce, which
indeed in the above confuses the reader about what his relationship was
at the time with Patsy. Were they separated, living in different
residences? Was Cathy staying with Patsy? Why would she call him at
that specific time? And where was Cathy staying when he hid from her
when their paths crossed in the previous January?

Were these lines written at the moment of the first draft? Or is their
present tense a rhetorical recreation of a nostalgic moment, a narrative
reconstruction, and not a momentary expression of the author's thoughts
at a moment of composition? In *Yonder* his lyrical outbursts have an air
of poetic license. He speaks out in direct address to wife, daughter, and
mother. The revealed composition date of some sections of *Yonder*
extend beyond his parents' death into 1988 and even beyond into 1989.
Why does he jerk his reader around with so much smoke and mirrors
about what was happening? Why should he think we do not care? And
finally it should take us somewhat aback when he denies that he is an
alcoholic. How much is he into denial? Or is he being sardonic? How
authentic are all his lamentations about loss and despair when he says
that he was confused in his own mind?

Authenticity, I suggest, reveals itself in a text when an author
demonstrates to a reader an ability to come to terms with self.
Authenticity is displayed to a reader when the author shows a careful
consideration of her motives and attitudes, reasons for commitments,
and reasons for beliefs. Authenticity displays a voice that seeks and
strives rigorously and conscientiously to achieve awareness of its own
actions and behaviors.

Corder is defensive, even apologetic, about why he didn't take note of
his children. He questions his memories of places, people, and changes.
He doubts who he has been. He reminisces about his historical and
cultural place, his origins in Jayton, Texas. His search for home, his
search for a sense of place, is a search into his own psyche. His nostalgic

recollections and researches into the history of West Texas are searches for his own identity. But I suggest that such searches are not the best way to find oneself. For the most part, we discover ourselves through others, in the commitments we have made with them. We discover it in our acceptance of changing opportunities and circumstances. An examination of one's life is much more authentic, I suggest, if we look for evidence that questions our self-conceptions and future commitments. How much am I truly committed to my professional aspirations? Do I really love those that I claim to love? At times he accepts his mother's past definition of himself, but he ultimately rebels against it.

My own differences with Corder are over issues about voice, persona in writing, and the rhetorical purposes served by expressive and narrative discourse. To question one's own personal identity, to fail to find it, to lament not having it, to me is an epistemic failure of sorts. And it is also a kind of moral failure. For me the discovery of who we have been is in large part found in an examination of our past acceptances, acknowledgements, and commitments. It is found in our personal narratives of why we made the choices and changes in our lives that we did. And that we make changes in our loyalties and beliefs is how we grow up. That we make changes is in part what it is to be educated. Surprisingly, Corder discusses little about his college and graduate education. He dismisses it all too easily. Certainly his education and scholarship made him into a different person. He gives no credit to any of his teachers.

To discover our identity, which for me is finding our center of decision, we need to examine both the logical and imaginative processes necessary to conceptualize and redefine our acceptances of, and commitments to, alternative paths of life. Certainly, as Corder emphasizes, our place, our home, our culture is focal in these reconceptions and redefinitions of acceptances and commitments. In part, defining personal identity takes place in our critical rejection of old values and beliefs. An authentic self certainly needs constantly to reexamine one's loyalties and current situational and problematic circumstances. But most important, genuine authenticity is to be discovered in the formation of ideals about the future person we choose to become. We locate it in an idealized self.

It has always intrigued me that so many want to be somebody without being anybody. They seek to be somebody without change. How can someone want to be educated without expecting to change as a human being? People fail to see that being somebody is something we need to achieve. Personal identity involves a search for the future. It is not found by simply dwelling on the past. An authentic person displays a cognitive and emotional search for a future self. There needs be some hope for a future in a search for self.

Suicidal people, depressed people, cease to think about the future. They make no plans for it. Suicide is an act of despair about hope for living. An Alzheimer's self, on the other hand, is a self that has no past and thus has no concept of an ongoing self, a self looking forward to a future. And the quest for self-knowledge that is nostalgic, such as goes on in *Yonder,* fails to recognize the ground of a past self and its influences as they enter into a new self-actualizing self.

I find it puzzling that Corder did not discover more about himself and what he was all about in therapy, which he surely must have undergone in psychiatric treatment. He does not mention trying to come to terms with himself and his family in his therapy. He did not talk about any changes in his lifestyle. His hospitalizations and treatment seem not to have contributed to self-understanding. His meditations over his Bloody Mary's and cheap white wine are certainly not quests for self-knowledge. What is one to make of his many attempts to escape into triviality: his escapes into his lists, maps, vegetable gardens, old movies, old airplanes, western myths and legends, and into the historical tales of West Texas. Are these serious and authentic interests?

What we demand of an authentic person is evidence that his or her self-examination be rigorous and thorough. We do not see in the narrative of *Yonder* any alteration of certain habitual behaviors and attitudes that almost certainly must have contributed to Corder's psychological depression. He continued to drink. He was obsessed with smoking tobacco. He continued to escape into isolation. In the closing sections of *Yonder,* the reader has little hope that he will escape the box that he has psychologically created for himself. He is always searching into diversions tangential to his profession. I never see in Corder the scholarly hunger that drove the productivity of a Kinneavy. And it always puzzled me why viable scholars such as Corder would take up administrative work at a university. Even if one is good at it, as he probably was, does it serve self-interest or enhance self-worth and personal integrity to do such work?

For me Corder's laments in *Yonder* are an elegy, not only for the death of his extended family but for the loss of his own family. He has written an elegy that laments his vanishment. And now I read it as an elegy for his own death. He writes in *Yonder* as if he had little hope for anything remaining of lasting value in life. As a memoir and as meditation, I find *Yonder* an expression of surrender. I find in it no commitment and no dedication to his profession in rhetoric.

The persona he develops in *Yonder* is open to suspicion. What adds to this suspicion at times is the lyricism of his lamenting voice. It appears exaggerated and embroidered with ornamental literary effects. The literary style of *Yonder* tends to shape a voice that has the air of literary fiction. How fictive is it? And that it may be partly fictive should bring into question some of the views Corder has on ethos and on what is our ethical responsibility in writing. How authentic can any writer be, given Corder's own personal views about himself? Sincerity and good faith, I suggest, are not enough for any writer. There is, I suggest, a rhetorical demand made of any author to give evidence of personal authenticity to critical readers. Faith in the sincerity of our expressive voices is not enough.

How much has Corder, in his writing, aimed at achieving literary quality? Much of *Yonder* has the tone, sonorities, and style of lyric poetry. There is wordplay, oxymoron, synecdoche, contrastive antinomies, paradox, assertion with simultaneous denial, questions, personal address, and above all numerous poetic refrains and poetic echoes in literary allusions. *Yonder* strives for a kind of poetic prosody. It is difficult for me when I attend to the rhythms of Corder's syntax to think of *Yonder* other than as having a literary motive and intention. He only minimally argues for the asserted opinions. Noteworthy in *Yonder*, the authoritative voice of his pedagogical works simply vanishes.

Corder's pessimism is without any ultimate consolation except escape into the personal annihilation of death. He sometimes speaks of epiphanies in *Yonder*, but there is nothing transcendental about them. There is no divine revelation in them. They are only momentary exaltations about beauty found in landscapes and music. There is no religious consolation in *Yonder*. There is only a mythic-poetic rendering of his Southern Baptist roots with Presbyterian conceptions of hell's fire and damnation, in which he speaks of himself as sinner, as one damned.

To whom is he confessing? Possibly, it occurred to me, he could have written this book for his children or even to his first wife, which makes the dedication to a second one somewhat ironic. So much allusion to dates and events known only to his own family suggests that the book in part was written as an apology to them. Why so many detailed descriptions of trips that he made with family in a car? They seem to be mere personal reminders to others about cherished intimacies.

But many sections of *Yonder* appear to be simply attempts at self-discovery. He most likely wrote these essays as exercises to come to terms with himself. But I find little in it that amounts to an imaginative reconstruction of a self that seeks achievements. There is little in *Yonder* that shows any allegiance to any overall cultural reform or political movement. Rather, in general, he simply laments the loss of a sense of place, a loss of family, and a loss of all faith. Surely he did not expect sympathetic understanding from readers with so little positive resolution and dedication. Surely he must have questioned whether he could find an audience for such exclamatory and personal lamentations of despair.

Corder's views on the rhetorical strategic importance of managing voice and persona are well documented in his pedagogical texts, especially in his college reader *Finding a Voice*. His focused discussion on the importance of voice in rhetoric should give us pause about any text such as *Yonder* by Corder. We should expect his voice and persona to be painstakingly managed and crafted. Has he in his voice fashioned an ideal of himself? Or has he created an image of himself that he wants his audiences to entertain? Is he selling himself? Or has he created a somewhat fictive persona that he wants to use for certain rhetorically strategic effects that leave intended contextual implications for his readers?

In looking at *Yonder* from the viewpoint of rhetorical criticism, what surprises me is the amount of evidence spread throughout the text yielding highly probable explanations for events that he supposedly chose to keep secret. Could he have deliberately left so much evidence implying what happened? Has he hidden his explanations from us simply to provoke our curiosity into drawing implications about the truth of what he seemingly wishes to hide?

From dates in his text, I find it easy to frame accurately the time lines of his life despite all the nonlinear digressions and descriptions found in his personal narratives. It has been easy for me to model his life and experiences on my own time lines since they are so much parallel. My life has bracketed Corder's. I was six years older. I was one of those World War II "fly-boys" that he spoke enviously about. I flew in training missions back and forth over the West Texas that he so nostalgically talks about. It is easy for me to follow most of the cultural allusions and the local color that he describes. During the war I made an exciting exit from Texas in a taxi with four fellow officers down the very same highway from San Angelo to Fort Worth that he describes in such cherished detail. The only thing alien to me in *Yonder* is the contemporary poetry and essays he cites. I too have raised okra and black-eyed peas.

I saw the backside of the Depression in a rural area. I went through many of Corder's traumas. I have been in a West Texas dust storm. I struggled in graduate school with children. I too was divorced and was alienated from my children. I too pursued an academic career at the expense of almost everything else. And I too remarried. My parallel interests in rhetorical studies provide me with a critical framework to read *Yonder* empathetically. I have no doubts about his sincerity, but it is his authenticity I question. But despite the nihilism in his lament in *Yonder,* I have always imaged Corder as essentially a man who chose to be moral.

Yet, he wants to cry out that he was selfish. In his writing he has always been very sensitive about hurting and injuring others. And that he has hurt others is what in *Yonder* leads to so much expression of guilt. But I do not think that he truly believes, with Augustine, that a predestined evil strain in him damned him. Breaking commitments, reaching compromises, is what morality is all about. We do not know from Corder's outcries how to empathize with his lament, for he has not personally engaged us in his confession of sin. He has not been open enough with us to satisfy our curiosity about him.

The person-narrator of the memoir format of *Yonder* necessarily brings ethos and persona into play. What makes *Yonder* good writing is its literary qualities, not the cognitive substance of his rhetorical views, nor the cognitive worth of his cultural criticism. And that brings me back to Kinneavy's writing style. Certainly Kinneavy's writing does not have the voice and persona management that we find in Corder. And insomuch as flow of sentences and clarity of style are part of voice and persona management, Kinneavy's thought is too excessively and theoretically abstract. It is analytic, definitionally precise, qualified with too many distinctions and quantifiable limitations. It does not allow his voice to play much of a part in the exposition of his thoughts. Nor does the complications of thought and refined distinctions in Kinneavy make for easily syntactically readable prose. Kinneavy's third-person authoritative voice is not one that invites empathy or attention to sonorities of language. The fact is that the linguistic patterns necessary to express Kinneavy's thought are at times almost baroque and not easily read.

What I question in the end is, should teachers of writing look for models of good writing in English text-readers, such as edited by Corder, where what is included is judged by literary standards? As an outsider I find it a serious problem for teachers of composition with literary backgrounds not to be seduced by their own literary standards of good writing. How much has Corder in his sonorities and artful wordplay failed to face the practical problems of argumentative and critical rhetoric? Why should our self-concept be under attack, as he maintains, when we engage in such rhetoric? Why are argument and explanation necessarily always personal?

In *Yonder* Corder found it somewhat disconcerting that he was asked to discuss the essay in a writing workshop when someone else was asked to talk about creative writing. That it happened tells you something about Corder's conception of himself as a teacher of writing and his conception of invention. How different is invention in literary and poetic production from invention in argumentative, expository, and critical essays? And how different and important is voice management in the two different sorts of writing?

WORKS CITED

Corder, Jim W. *Finding a Voice*. Glenview, IL: Scott, Foresman, 1973.

—. "When (Do I/Shall I/May I/Must I/Is it Appropriate for me to) (Say No to/Deny/Resist/Repudiate/Attack/Alter) Any (Poem/Poet/Other /Piece of the World) for My Sake?" *Rhetoric Society Quarterly* 18 (Summer 1988): 49-68.

—. *Yonder: Life on the Far Side of Change*. Athens: U of Georgia P, 1992.

Hirsch, E. D., Jr. *The Philosophy of Composition*. Chicago: U of Chicago P, 1977.

III

Parallels, Extensions, and Applications

8

A Call for Comity

Theresa Enos
University of Arizona

☙Cultural historians are busy charting and critiquing the national erosion of comity—mutual courtesy, civility—in the public sphere. Media examples of the coarsening of our society, the erosion of manners, and the rise of incivility (speech or action that is disrespectful or rude) have in recent years increased sharply. "Sifting the shards of America's fractured taste and manner," as conservative columnist George Will puts it, cultural historians are suggesting that the "casual coarseness" and incivility so evident today in public action and discourse is symptomatic of the "New Economy" ("Coarsening of Society Product of New Economy"). Our forebears, at times, may have been ignorant of manners, but they were not hostile to them; today many are convinced that we are both ignorant and hostile to manners and matters of civility.

Cultural critics are blaming nearly everything for "our incivility crisis": the automobile, telephone solicitation, the microwave oven, the birth control pill, radio (the Sterns, Imuses, and "other heroes of 'shock radio'"), television (just about anything, but especially shows such as *The McLaughlin Report* and *The Capital Gang*), and the rise of cyberspace (Carter 193). And we've become high-tech barbarians according to newspapers, magazines, and TV. High-tech innovations, many think, have helped hasten the deterioration of manners and have made us become desensitized to each other. Who can have escaped cell-phone users talking loudly in public spaces? Silicon-Valley types, our young dot.comers, proselytizing downloading, inputting, hacking, spamming? A world on fast forward . . .

And recent charges against the entertainment industry leave little doubt for many why popular culture looks so much like raw sewage. Questions are being raised whether we want the industry to sell garbage to our children, to feed them a steady diet of games and videos that glorify violence, cheapen human sexuality, reinforce the worst gender-stereotyping imaginable, and in the process render civil behavior a Victorian concept not worth mentioning, much less practicing.

But along with this cultural critique, we also hear about attempts to legislate civility. Some public schools are requiring teachers to teach not only math but also manners and morals. "Courtesy Equals Respect" is an educational program made law in Louisiana's public schools in 1999.

Twenty-eight states now require or encourage character education, making ethics part of the K-12 curriculum. Other signs of a culture rebirth include required school uniforms, the teaching of table etiquette, and even the return of the cotillion by parents eager to instill some manners in their children.

In the political sphere, forty-three candidates for federal office by 2000 had signed the Interfaith Alliance's Framework for Civility, vowing to reject personal attacks. The Alliance, representing many religious denominations, began asking politicians to keep a civil tongue following the mean-spiritedness that buffeted the nation during President Clinton's impeachment. To get elected officials to act more seemly in the political arena, the Alliance asks them to be models for integrity, fairness, and respect. One of George W. Bush's promises in the 2000 presidential race was to restore civility and bring a new tone of collegiality, vowing to "end the arms race of anger" in American politics. Yet after the Supreme Court's decision to end the vote recount in Florida, Justice Clarence Thomas speaking at the American Enterprise Institute in Washington, DC, apparently rejected Bush's call for compromise and harmony, saying "Today there is much talk about moderation," but there is an "overemphasis on civility." "Civility cannot be a governing principle of citizenship or leadership," Thomas said, adding: "Though the war in which we are engaged is cultural, not civil, one should not let principle be cannibalized" (Dowd B7).

Higher education has not exempted itself from the sense that incivility is the *sin du jour*; we are familiar with the unsuccessful attempts to restrict or ban the barrage of offensive speech from campuses, and the *Chronicle of Higher Education* has featured several articles on the breakdown of decorum in classrooms (see, for example, Schneider). In her research on incivility in public-school classrooms, Carbone presents evidence that large classes are part of the cause of abominable behavior, and she thinks such incivility is "caused by the same mind-set that allows otherwise polite individuals to gesture rudely at other motorists in a traffic jam or shout obscenities at a referee at a crowded sporting event" (37). The anonymity and impersonal nature of a large class can inspire students to behavior they would never dream of exhibiting in their small classes.

On my own university's English faculty listserv in spring 2000, a discussion on salary equity became overheated; personal attacks began. Some were aghast, especially when others jumped in, excoriating their own colleagues for what they considered to be substandard scholarship, both in quality and quantity. Such acts of incivility well illustrate Dr. Johnson's remark at a dinner party that once civility has been breached, we can be as rude as we like (*Boswell* 972).

The craving for immediacy, the journalistic criticism "of-the-moment-at-the-moment" mindset that Henry James once decried (qtd. in Weaver, *Ideas Have Consequences* 111), and the loss of any distinction between the private and the public, cultural critics suggest, are at least partly to blame for the real or perceived coarsening of society. Incivility, as Wayne Booth has told us, is itself a form of communication, and what it communicates is that something is wrong with a relationship. What I want to explore is how compressed time works against a spacious discourse that requires a sense of expanded time, space, and self-reflexiveness. Along this exploratory route, I present some of Deborah Tannen's ideas on the "argument culture" in today's public sphere; then I'll go on to trace the history of manners and offer some of Aristotle's ideas on *civis* and *civitas*. Along my path I will work with some rhetorical architects—Richard Weaver, Kenneth Burke, and Jürgen Habermas—in order get to the promised "call for comity," a call I take from Jim Corder's own rhetoric that foregrounds ethos.

Am I a retrograde here? Perhaps—but I'm not quite of the Great Gray Legion who came to maturity during a time when manners, and civility, were so important. Our time is noisier, more complex, more compressed; "We must pile time into . . . discourse," Jim Corder urges, to help us make sense of a world that doesn't want time in its discourses ("Argument" 31). I may seem retro, too, in my choice of rhetorical architects from whom to cull ideas (except for Tannen, I realize, they can and often are seen as retros—especially Weaver). So I'll plead guilty as-surely-charged with retrograde ideas, but then, in my own "partial defense" of some balancing rhetoric, I know that others are beginning to question whether we should situate ourselves only in the postmodernist world of rhetoric and theory, excluding other times, spaces, and rhetorics. I am heartened by John Trimbur's recent essay on the "partial" defense of modernism wherein he critiques pomo's "new good decentered nomadic textual self" and, wants (partially?) to raise the author from the pomo-dug grave (284). What Trimbur argues against is our tendency for overcorrection (specifically in his essay, "over-correcting" the Romantic notion of the function of an author). I agree with Trimbur that we might well consider the "corrosive effects of postmodernism" and that we should be cautious about burying all of our old theories about discourse in trying to raise ever-higher postmodernist edifices (297). This chapter is my attempt to begin what I hope will be an ongoing dialogue about how we might make our rhetorics more spacious.

One of the few who have written on rhetoric and incivility in the public sphere is Rolf Norgaard; in "The Rhetoric of Civility and the Fate of Argument," he presents a note of caution lest we all jump to agree with the pundits that "today's social and political discourse has deteriorated to levels of discord that extend beyond rudeness to acts of intolerance and disrespect" (247). He argues that the cultural critics are ignoring "the rhetorical dimensions of civility" and that when the discussion extends beyond manners and etiquette, it "generally takes off in the direction of political philosophy" (248). Instead of defining civility as Aristotle does, as a contextual virtue, as a means between extremes, which would "rob" our cultural critics of their "most prized commodity: moral certitude" (252), Norgaard argues, "we should be positioning the concept in terms of rhetoric's roots in the *polis* so that the link between rhetoric and civility is readily apparent, especially in relation to epideictic discourse and the fact that a "rhetoric of civility requires of us all a heightened level of self-reflection—one that is missing in today's appeals" (248). I hear this need for caution.

THE CULTURE OF CRITIQUE

I want to discuss some of Deborah Tannen's views in her most recent book, *The Argument Culture*, not so much because they will be new to those of us in rhetoric and composition but because she is a well-read cultural critic. The argument culture, Tannen says, pits us against one another: The best way to begin an essay is to attack another, the best way to show your mind is really working is to criticize. Criticism, attack, opposition, these are all part of incivility. "Conflict and opposition are as necessary as cooperation and agreement, but the scale is off balance, with conflict and opposition overweighted" (4). We don't much consider building time and space; when we're in an argument with someone, our goal is not to listen and understand but to use every available tactic we can find, "including distorting what your opponent just said—in order to win the argument" (5). Take academic papers, Tannen says—and here most of us would readily agree: "The standard way of writing an academic paper is to position your work in opposition to someone else's, whom you prove wrong. This creates a *need* to make others wrong, which is quite a different matter from reading something with an open mind and discovering that you disagree with it" (268-69).

Tannen also thinks that our high-tech world prevents us from a spacious discourse. E-mail is particularly interesting in that it is more personal than the phone; we open up more, much like writing in a journal, because we tend to feel alone in the process of thinking and writing down words. But e-mail, wherein we too often just toss out verbalizations, also can aggravate aggression. Just as e-mail and the Internet are creating networks of human connections unthinkable in the very recent past, "at the same time that technologically enhanced communication enables previously impossible loving contact, it also enhances hostile and distressing communication" (139). Flaming (vituperative messages that verbally attack) is a result, Tannen says, "of the anonymity not only of the sender but also of the receiver. It is easier to feel and express hostility against someone far removed whom you do not know personally"—similar to incidents of road rage (239).

Because present-day technology provides both speed and anonymity, it too easily sparks hostility and attack. Tannen cites Walter Ong and Robert Oliver in her suggestions on how we might move from the more hostile debate to the more open dialogue. She quotes Ong, who suggests that if debate seems the only path to insight, we should take note of the Chinese approach: Disputation was rejected in ancient China as "incompatible with the decorum and harmony cultivated by the true sage" (258). And the more spacious exposition is in Chinese rhetoric the "preferred mode of rhetoric" . . . rather than argument. The aim was to "enlighten an inquirer," not to "overwhelm an opponent" (qtd. in Tannen 258).

The argument culture today lets us too often cancel each other out; we make it even harder to "know" what is "true" when we search for something to refute rather than opening up and taking a more indirect route—and taking the time to listen (a point I'll explore more in depth later when I discuss Corder's rhetoric). "If you limit your view of a problem to choosing between two sides," you are not framing the issue in the most constructive way; thus "you inevitably reject much that is true, and you narrow your field of vision to the limits of those two sides, making it unlikely you'll pull back, widen your field of vision, and discover the paradigm shift that will permit truly new understanding" (Tannen 290). Tannen doesn't advocate that we have to give up conflict and criticism altogether when we move toward a more expansive view of argument; rather, "we can develop more varied—and more constructive— ways of expressing opposition and negotiating disagreement," especially through working toward an overlapping that can enlighten both parties (290). What we find is that we can focus on commonalities, finding "resonances" between speakers; "let the resonances speak for themselves," she pleads (7).

> *If you would civil your land, first you should civil your speech.*
> —W. H. Auden (*The Dyer's Hand*)

What do we mean by *civility*? To Tannen civility "suggests a superficial, pinky-in-the-air veneer of politeness spread thin over human relations" (3). This is an easy-enough definition when we can still so vividly picture John McCain's strained concession appearance with George W. Bush or Hillary Clinton and Rick Lazio's stiff handshake matched by their brittle, insincere smiles. I would suggest that this might be a better illustration of manners, certainly part of civility. We might better define *civility*, or *comity* (*comity* is a more inclusive concept than *civility*) as "the attitude and ethos that distinguish the politics of a civil society, i.e., a solicitude for the interest of the whole society, a concern for the common good," which includes a culturally ingrained willingness to tolerate behavior that is in some degree offensive (Shils 1). The space between the state and the individual is what we call civil society.

Citizenship, civility, civilization are cognates from the *cîvis* ("citizen") and *civitas* ("city"), Latin equivalents of the Greek *polis* (interestingly, the word *polite* has its roots in *polis*). "In a liberal democracy," Edward C. Banfield notes, "the citizen is one who has the right to vote and who therefore shares the obligation to protect the rights pertaining to the private sphere. Citizenship implies a sense of shared responsibility" (x), and it "implies membership in a community defined by a common substantive end, more comprehensive, more dignified, more authoritative than the particular ends of private individuals (Orwin

75). Pericles' "Funeral Oration" might best clarify the distinction between citizenship and civility; in his praise of Athens for its remarkably open society, he also defines the concept of civility both in the public sphere and in the private sphere. Orwin explicates thusly:

> Whereas its arch-rival Sparta depends for its stability on constraint and coercion, Athens encourages the free flowering of human perfection. This perfection encompasses alike the distinctive excellence of the citizen and that of the human being as such. Of greatest relevance to us is that Athens as Pericles presents it is distinguished by its amazing freedom not only in public matters but in private ones. (77)

About private freedom Pericles exhorts:

> And not only in our public life are we liberal, but also as regards our freedom from suspicion of one another in the pursuits of every-day life; for we do not feel resentment at our neighbour if he does as he likes, nor yet do we put on sour looks which, though harmless, are painful to behold. But while we thus avoid giving offence in our private intercourse, in our public life we are restrained from lawlessness chiefly through reverent fear, for we render obedience to those in authority and to the laws, and especially to those laws which are ordained for the succour of the oppressed and those which, though unwritten, bring upon the transgressor a disgrace which all men recognize. (Thucydides II.37)

The public sphere takes precedence over the private; citizenship takes precedence over civility, "in which the diversity of the society (and the value placed on diversity) is trivial in comparison with the demands of citizenship considered as an overarching unity" (77). Comity and civility are qualities that imply the restraint of anger directed toward another, but it is not the same thing as warmth and indeed implies a certain coolness, as Kesler points out: "Civility helps to cool the too-hot passions of citizenship" (57). Stephen Carter in *Civility: Manners, Morals, and the Etiquette of Democracy* goes even further, positing that "Civility has two parts: generosity, even when it is costly, and trust, even when there is risk" (62).

If, then, we can view civility, overall, as something positive, certainly more than Tannen's sense of it as a cool, rather-distant, "raised pinky" kind of behavior, we can now work with Aristotle's concept of *homonoia* in the public sphere, which in its translation of *concord* and in Aristotle's context, is a part of comity, most specifically in that it "includes any sort of kindly feeling, even that existing between business associates, or fellow-citizens" (*Ethics*, fn 450). Aristotle points out that human beings have a natural friendship for each other:

Even when traveling abroad one can observe that a natural affinity and friendship exist between man and man universally. Moreover, friendship appears to be the bond of the state; and lawgivers seem to set more store by it than they do by justice, for to promote concord, which seems akin to friendship, is their chief aim, while faction ["civil conflict"], which is enmity, is what they are most anxious to banish. And if men are friends, there is no need of justice between them; whereas merely to be just is not enough—a feeling of friendship also is necessary. Indeed the highest form of justice seems to have an element of friendly feeling in it. (viii.i.3-4)

Here is Richard Leo Enos's helpful gloss:

The virtue that Aristotle is discussing in the beginning of Book 8 is "friendship" (philia). Humans have a clear trait to bond together as a race (8.1.3) in what Aristotle calls "philanthropic" (philanthropous). Philia can help a state (communities) bond together, which promotes concord (homonoia). For Aristotle, homonoia is the equivalent to friendship and the objective of lawgivers. In fact, lawgivers may see this as more valuable than even justice. (E-mail Oct. 5, 2000)

When citizens are civil to one another, despite their political disagreement, they reveal that these disagreements are less important than their resolution to remain fellow citizens. *Homonoia* certainly is linked to *virtù* and, by extension, to ethos. There are two kinds of virtue: intellectual and moral. "Virtue of thought arises and grows mostly from teaching, and hence needs experience and time. Virtue of character [*êthos*] results from habit [*ethos*]" (*Ethics,* Trans. Irwin 1103a14.17). To Aristotle it was clear that none of the moral virtues is engendered in us by nature, for no natural property can be altered by habit (1103a19). We acquire the virtues by practicing them: "It is by taking part in transactions with our fellow-men that some of us become just and others unjust. And character qualities that are acquired through practice can be destroyed by excess or deficiency" (1104a14.17). This seems consistent with what we know about the development of character in people; Aristotle's argument is that moral virtues, unlike the intellectual ones, are the product of the regular repetitions of the right actions. We are habituated to temperate and moderate behavior by routinely acting in temperate and moderate ways. Virtue has the characteristic of hitting the mean and, importantly, knowing what that mean is—the best amount and in the right manner toward the right people for the right purpose— this is the mark of virtue. Virtue is a "settled *disposition* of the mind determining the choice of actions and emotions, consisting essentially in the *observance* of the mean" (1106b; emphasis added). Virtue is linked with morality, or at least, as Aristotle is suggesting, that all our actions must meet the test of morality, that we should practice our ability to do what is right rather than what we desire. Civil people to Aristotle meant those "who do not need extrinsic reasons to be moral; they act for the sake of the virtues themselves, guided by practical wisdom" (Kesler 66). Thus comity can be seen as a form of magnanimity, Kesler further explaining Aristotle's view: "Such greatness of soul [comity] accepts external honors as the highest tribute that can be paid it, but regards all such popular offerings as vastly inferior to its own sense of dignity and propriety" (61).

We acquire virtues, then, through choice and training so that they become a stable part of a mature person's traits. In the *Nicomachean Ethics*, Aristotle gives *phronésis*, one of the intellectual virtues, special attention. *Phronésis*, practical wisdom, makes it possible to internalize and practice other virtues. It is through deliberative civic conduct that *virtù* is cultivated. Aristotle's treatment of *phronésis*, as Farrell explains it, is "rhetoric properly practiced, which allows the rhetorician to become a more accountable moral agent . . . literally to cultivate *phronésis*" (98).

Virtue as something that can be taught and practiced until it becomes habitual can be traced back further than Aristotle, Jarrett and Reynolds say, as far as Plato's *Protagoras,* where Protagoras outlines the "process of socialization from family to school to public service in the *polis"* (45). Virtue can be taught, Jarrett and Reynolds agree; furthermore, "rhetoric can, indeed must, articulate culturally specific ethical codes" (45). In ancient times there were lifelong programs to develop excellence through education; the public self constructed in this way is continuous with the private. Aristotle tells us that anyone can get angry, but what we should strive for, when necessary, is disciplined anger; to be angry with the right person, and to the right degree, and for the right purpose, and in the right way—that is disciplined anger.

Cicero in his *Orator* distinguishes between *ethikon* (ethos) and *pathetikon* (pathos), pathos being violent and hot, ethos being courteous and agreeable where authority springs from one's nature and character— and depends largely on virtue. Another of Cicero's concepts is explained by Richard Enos and Karen Schnakenberg: *dignitas*, a shaping characteristic, "a standard that went beyond meeting the immediate expectations of an audience to meeting those ideals that appealed to Romans as a community." Furthermore, "Cicero's notion of *dignitas* reveals his view of ethos, that the composition of rhetoric must meet not only the standards of an immediate audience but the social standards of Rome as a community and a culture" (201). One is a Roman first, a Marcus or Lucius second. A Roman citizen "is an individual, with his own needs, interests, concerns, affections, only to the extent compatible with citizenship in the full and primary sense, and so only to a very limited degree" (Orwin 76).

During the Renaissance Machiavelli believed that *virtù*, in the sense of civic virtue, was a classical idea—rare but attainable. In bringing virtue back to its Latin root *vir*, meaning manliness or prowess, he wanted the concept to reflect taking events into one's own hands, shaping them according to one's will rather than the more pious meaning the word had taken on beyond the classical period. Later, Enlightenment philosophers "expected civic virtue to arise as the predictable, though not the inevitable or unintended, consequence of living in a properly constituted state" (Auspitz 18). By the nineteenth century, civic virtue was regarded not only as a worthy goal but as an urgent practical necessity; thus the ideal had been enlarged to include conduct that most rather than a very few citizens might be imagined capable of.

According to the Victorians, a good citizen was to reflect moral values rather than the older meaning of *virtù*, the veneer of Victorian good manners replacing the classical ideal of virtue. Victorian ladies and gentlemen were taught the art of conversation. Carter points out that if the "art of Victorian conversation sometimes involved a skill at evading discussion of difficult or embarrassing issues, it also carried a strong current of civility" (138).

I suggest that comity too observes the Aristotelian mean and that it is closer to the concept of virtue, to ethos—and, I'd hope, moral (though we are at risk to use the term today), than it is to what we seem to mean by values. And there has been much talk about what we mean today by *values*. George Will professes that we should stop talking about values and, instead, as our Founders did, talk about virtues. Will points out that in the political arena, it's not the singular *value*, used as a verb meaning to esteem, but the plural noun, *values*, "denoting beliefs or attitudes" ("Teach Virtues" B7). "Values" is democratic; "unlike virtues, everyone has lots of values, as many as they choose." Hitler, Will says, had "scads of values. George Washington had virtues"—one never hears anyone talk about Washington's "values." Surely drawing from Aristotle, Will posits that virtues are habits, difficult to develop, and therefore not equally accessible to all. "Speaking of virtues rather than values is elitist, offensive to democracy's egalitarian, leveling ethos"—which is "why talk of virtues" in the way De Tocqueville and the Marquis de Lafayette did, "should be revived." De Tocqueville "noted that although much is gained by replacing aristocratic with democratic institutions and suppositions, something valuable is often lost—the ability to recognize, and the hunger to honor, hierarchies of achievement and character" ("Teach Virtues" B7). Thus the term *values* often is a perfect nondescript product of the nonjudgmental among us whereas *virtue* defines what is morally excellent, or good. And *moral* implies a will in making judgments as to the good or evil of our actions.

THE CIVILIZING PROCESS AND THE
HISTORY OF MANNERS

The ancient *polis*, and many modern political societies that are not really based on the primacy of the rights of the individual, start with the community, deriving the duties of citizenship and what they call rights from the nature and needs of the community. In our own culture, there's the sense that the concept of both *civic*, meaning "citizenship," and *civil*, meaning "civilized," has eroded. The history and practice of manners show how the civilizing process is a necessary part of the communal and public sphere. The "rules of civility" began to develop

> in Europe at the same time that the church began to yield its authority to the nation-state, and they served the important function of teaching the importance of gaining control over our instincts. People could no longer kill others with whom they had grudges, and they could no longer urinate where they chose; the relationship between the two is that both are natural urges, and both had to be cabined, if only for the society to survive. But there was a moral aspect as well: the ability to discipline the desires of the animal self. (Carter 277-78)

Norbert Elias, the Swiss sociologist who did the first important scholarship on the history of civility, points out that "The standard of what society demands and prohibits changes; in conjunction with this, the threshold of socially instilled displeasure and fears moves; and the question of sociogenic fears thus emerges as one of the central problems of the civilizing process" (xii-xiii). Elias identifies a pattern of civility that leads to psychological constraints on impulse—incivility is primitive; civility, civilized. In *The Descent of Manners*, Andrew St. George also traces the history of manners, focusing on behavior books, a genre that he traces back to Erasmus. "Manners," St. George says, "are much more than a miscellaneous collection of changing social rules; there is a link between everyday norms of behaviour and the overall values of a society" (2). Both Elias and St. George make a similar argument: In the progress of civilization, primarily from the twelfth through the nineteenth centuries, there is a process of increasing inhibition. The sixteenth century thought that society could preserve itself by courtesy and humanity by certain standards of behavior accepted and practiced.

In the Middle Ages, *civilité*, with its echo of chivalry, acquired its meaning for Western culture, but the concept of *civilité* seems stamped in the sixteenth century with the first treatise on civility, Erasmus's *De civilitate morum puerilium (On Civility in Children)* (1530). *De civilitate* is very much in the tradition of such medieval manuals on the usages of courtesy and the knightly conception of duty. Erasmus's behavior book had an enormous circulation, going through edition after edition—130 editions, 13 of them in the eighteenth century. There were many translations—and imitations. Elias says that "Erasmus gave new sharpness and impetus to the long-established and commonplace word *civilitas*," expressing it as "something that met a social need of the time" (43). The subject matter is simple: the behavior of people in society. Erasmus speaks of the way people look ("a wide-eyed look is a sign of stupidity," "staring a sign of inertia" [qtd. in Elias 44]), and outward behavior—bodily carriage, gestures, dress, facial expressions—is an expression of the inner.

The language is direct:

> The nostrils should be free from any filthy collection of mucus,
> as this is disgusting (the philosopher Socrates was reproached
> for that failing too) [earlier in the passage, Erasmus had said
> that Socrates had a bad habit of "gaping as if in
> astonishment"]. It is boorish to wipe one's nose on one's cap or
> clothing; to do so on one's sleeve or forearm is for fishmongers,
> and it is not much better to wipe it with one's hand, if you then
> smear the discharge on your clothing. The polite way is to catch
> the matter from the nose in a handkerchief, and this should be
> done by turning away slightly if decent people are present. If,
> in clearing your nose with two fingers, some matter falls on the
> ground, it should be immediately ground under foot. (274-75)

It is impolite to greet someone who is urinating or defecating; and "Let no one, whoever he may be, before, at, or after meals, early or late, foul the staircases, corridors, or closets with urine or other filth, but go to suitable, prescribed places for such relief" (qtd. in Elias 107). Furthermore, Erasmus says, a well-bred person should always avoid exposing, "save for natural reasons, the parts of the body which nature has invested with modesty" (277). And to those who teach that one should retain wind by compressing the belly, Erasmus says: "It is no part of good manners to bring illness upon yourself while striving to appear 'polite.' If you may withdraw, do so in private. But if not, then in the words of the old adage, let him cover the sound with a cough" (278). Striking to readers is that the "unabashed care and seriousness" with which behaviors are so publicly discussed and examined and that have subsequently become "highly private and strictly prohibited in society emphasizes the shift of the frontier of embarrassment" (Elias 107). Toward the book's end, Erasmus goes beyond this fixed pattern of "good behavior" and discusses conduct in a more comprehensive way. He speaks of *civilitas*, of courtesy: "If one of your comrades unknowingly gives offense . . . tell him so alone and say it kindly. That is civility" (289).

One of the more interesting arguments Elias makes is why there is a different relationship between earlier societies and people today. Americans might marvel at Europeans who took food with their fingers from the same dish (of course, a number of present-day cultures foreground communal eating from the same dish), not to mention other peculiarities. Elias says about the then and now differing social relationships:

> [T]his involves not only the level of clear, rational
> consciousness; their emotional life also had a different
> structure and character. . . . What was lacking in this *courtois*
> world, or at least had not been developed to the same degree,
> was the invisible wall of affects which seems now to rise
> between one human body and another, repelling, and
> separating. (69)

In other words, earlier cultures did not see the self, as we do, as being so radically different from another self. "Rather than our norms of individualism," Carter says, "they perceived themselves as part of a larger whole" (278). Carter doesn't argue for returning to a "world in which individual identity was subsumed within a larger and often brutal whole," even if we could, but "What we can do is try, within the limits of democracy, to construct a civility that will lead future generations to admire what we tried to do for civilization rather than condemn us for our barbarism" (278-79).

MANNERS AND CIVIL SOCIETY

The notion of civility as a process of self-betterment in the public spaces of the sixteenth and seventeenth centuries, St. George points out, "was reinvented by the mid-Victorians as politeness and correctness" (6)—thus the descent of manners from a civic duty based on the Aristotelian concept of virtue gained through habituation to a cool, rather distant, "the raised pinky" kind of behavior. For a set of manners to work, there has to be a set of beliefs that the culture takes for granted, and we can see this especially in midnineteenth-century England (much more so than in America where manners were being made up as the country defined itself). By the eighteenth century, St. George argues, in Britain courtesy gave way to etiquette, and by the midnineteenth century, "etiquette gave way to manners and became a class-based set of rules for admitting oneself to keeping others out" (7); "Manners," says St. George, "are social control self-imposed; and etiquette is class control exercised" (xiv). The polite conduct of the sixteenth century now becomes an argument for completeness. One should aspire to being socially and politically polished. One should aspire to a lifestyle of goodness, wisdom, and temperance of mind.

Manners amount to an unwritten social constitution, and St. George makes the point that because there appears to be no political bias behind the prescriptive, the "rules" they promoted took on the "vesture of natural law ... or from the viewpoint of some today, ideology in its most powerful form" (xiv). Good manners meant "communal integration and ingratiation" (7).

Appropriate conduct for the Victorians became a habit of mind—in conversation, in business, in one's creative work. Whereas the eighteenth-century rationalism of Locke recognized rules of engagement for conversation—"taste"—the mid-Victorians were concerned with how one conversed. The etiquette books of the midnineteenth century in English were "first to tackle the issue of how one conversed with a view to self-advancement," St. George says; "What could be more personal than conversing? What could be more political than conversing correctly? The mid-Victorians were the first to put conversation in service of a vision of society, whether for personal advancement, for religious purposes, or for political control" (48). The art of conversation during this time displayed in its forthrightness and surety the force of a national character as well as the personal; The "grand rhetoric of the assembly hall waited at the end of the enterprise which began in the drawing room" (77). And the belief in the primacy of conversation foregrounded the moral implications of communicating with others.

All of this, of course, was connected to the elocutionary movement. Edwin Drew's *The Elocutionist Annual for 1889* stressed that the "real beauty of elocution lay in its availability for practice, wherever one went a kind of portable gymnasium for self-improvement" (qtd. in St. George 77). Thus the concerns bearing on the practice of conversation pressed on the Victorians from several directions: "The writers of conversation manuals treated it as an art to be acquired; divines and teachers as a means of instruction; and elocutionists or public speakers as a model for delivery" (St. George 78). Even the 1851 Great Exhibition is a telling example of how mid-Victorian England built public spaces in order to make private spaces transparent. The Great Exhibition's best example of public architectural achievement was a home pretending to be a public thoroughfare; mid-Victorian England was the "last environment capable of building public spaces commensurate to the task of social cohesion and division, but chose the different, domestic route" (85).

By the midnineteenth century, the idea of the process of evolutionary change entered the Victorian consciousness; some saw the period as the Age of Transition, the change at first seen as a shadow, a shadow of ideas traveling upward—and manners downward. Meanwhile, attention was beginning to be focused on the culture and manners of America.

The American colonists, most of them coming from the peasant and working classes, didn't bring with them any tradition of polite behavior. But slowly in eighteenth-century America, standards of civility, similar to those in Britain, were enforced. "People were punished, usually in public, for scandalmongering, cursing, lying . . . 'finger-sticking' and making ugly faces" (Schlesinger 1). In the nineteenth century, de Tocqueville found the civility standards and manners pretty well accepted and practiced, though he found American manners "neither so tutored nor so uniform" as in his own country but "frequently more sincere" (qtd. in Schlesinger 26). No doubt de Tocqueville saw the concept of civility through the eyes of his great teacher, Rousseau—as a moral compass, pointing one to a genuine respect for the rights of all people in one's community.

Civility to Americans during this time was a matter of moral education, involving the shaping of young people's character through precepts, examples, exhortations—and shame liberally applied when the situation permitted it, or demanded it. With the subsequent rising of the American middle class, however, the practice of civility was not as closely identified with morality as it had been but was becoming separated from ethics, "taking on its modern meaning of a generalized pattern of behavior designed to lubricate social discourse" (Schlesinger 65). To some people fixed social rules have appeared a system of hypocrisy; to some, heartless formality; to others, an invidious expression of class distinctions. In Schlesinger's words: "To some, the purpose of civility is to enable one to seem a gentleperson without being one. To others, the precepts are intended to make one seem externally what one ought to be internally" (68).

The "descent" of manners from the nineteenth century through the twentieth century—and into the twenty-first—has helped lead to the present-day refrain, "a little civility, please." Has social discourse suffered from the "coarsening" of our society? Have we separated ourselves too far from ethics and morals? Most cultural critics think so. I don't purpose to try to answer such questions, although I cannot ignore them. What I do want to explore next is the compressed time and space in our culture that works against comity and cooperation. Along with this is the increasing recognition that the distinctions between *civic* and *civil*, between private and public spaces, has just about vanished. I turn to some twentieth-century theorists whose ideas might help us make that call for comity.

RHETORICAL ARCHITECTS: BUILDING TIME FOR COMITY

> *The corruption of man is followed by the corruption of language.*
>
> —Emerson ("Nature")

The theories and cultural observations of those I'm calling architects of time and space seem diverse; I hope to show, however, that all of them are concerned with the spaciousness of discourse, built in large part on reflexivity, that can lead to a shared sense of social harmony. This commonality is what links Richard Weaver, Kenneth Burke, Jürgen Habermas, and Jim Corder. All of Corder's writings show us how to spread out time and space whether he is speaking of the realm of invention, of the power of narrative over confrontation, voice, and, always, ethos.

Weaver's conservative statements on race, democracy, and education make one pause, but I believe we cannot neglect Weaver as a rhetorical architect. (For an opposing view, see Crowley.) I would argue that some of his thinking is useful as a counterstatement to some postmodernist thinking that too often leads us toward a tendency to overcorrect. Weaver decries our culture's scientistic orientation that values empirical "truths" while in other situations relativism seemingly runs amok—anyone's opinion is as valid as anyone else's. Burke and Habermas also blame science for the breakdown in cultural values and the resulting fragmentation of society. Weaver, never veering from his Platonism, believes that there is some stable and communal truth, some universal principle. In the late 1940s when Weaver began *Ideas Have Consequences*, he felt strongly that we were in an age of pretty bleary moral relativism: "This is another book about the dissolution of the West. . . . There is ground for declaring that modern man has become a moral idiot" (1). Weaver attributed this to the denial of logical realism and the denial that universals have a real existence, that is, the denial of everything transcending experience: simply, the denial of truth and the release of a relativism that has spiraled out of control. He looks to rhetoric to restore the habit of moral thought and action. For Plato, truth was a living thing, never wholly captured; today, however, Weaver says, "The more firmly an utterance is stereotyped, the more likely it is to win credit" (*Ideas* 95-96). Weaver uses as his primary metaphor the stereopticon, a kind of slide projector designed to allow one view to fade out while the next is fading in. He says the "function of this machine [is] to project selected pictures of life in the hope that what is seen will be imitated" (93). Weaver's "Great Stereopticon" can easily translate for us Plato's famous figure of the cave. The prisoners, as we recall, cannot perceive the truth because on the wall before them are only shadows of the real being projected behind them and that they cannot see.

The fragmentation of society has arisen, Weaver says, because those in the public sphere have no idea of how to

> persuade to communal activity people who no longer have the same ideas about the most fundamental things. In an age of shared belief, this problem does not exist, for there is a wide area of basic agreement, and dissent is viewed not as a claim to egotistic distinction but as a sort of excommunication. (*Ideas* 92)

In *The Ethics of Rhetoric*, Weaver says the way we habitually form an argument expresses our values; thus forms of argument characterize the user. Weaver describes and analyzes three kinds of argument that vary in building time and space into discourse: (a) definition, arguing about the nature of things; (b) similitude, making essential analogies; and (c) circumstance, arguing from a present circumstance. Weaver, as we know, valorizes the first—argument from definition—pointing out that arguing from circumstance is arguing from a sense of urgency, which admittedly has the power to move but is, ultimately, shortsighted. It is an "inferior source of argument, which reflects adversely upon any habitual user and generally punishes with failure" (*Ethics* 83). Since it is "grounded in the nature of a situation rather than the nature of things, its opposition will not be a dialectically opposed opposition" (83). When we discourse about the nature of things, and/or offer analogies, we are creating at least the opportunity for a spacious rhetoric because of the demand for reflection in this type of argument.

In "The Spaciousness of Old Rhetoric," Weaver examines a type of discourse where time and space are built in, an antiquarian one to be sure, the kind of oratorical event that our grandparents or great-grandparents would ride in wagons to hear on some festive occasion. The "pretentiousness" of this kind of rhetoric today might make us feel somewhat embarrassed. Nevertheless, Weaver says, that it can cause a sense of discomfort "as distinguished from indifference, suggests the presence of something interesting" (*Ethics* 164). Between the words—the actual speech—and all that it is meant to signify, there is "some space" that prevents "immediate realizations and references" (164). The "large and unexamined phrases" of this nineteenth-century oratory give it its grand style, the phrases having "resonance," and this resonance is what Weaver calls "spaciousness." "Instead of the single note . . . they are widths of sound and meaning; they tend to echo over broad areas and to call up generalized associations. This resonance is the interstice between what is said and the thing signified. In this way the generality of the phrase may be definitely linked with an effect" (169). Earlier I mentioned Weaver's recalling Henry James's description of journalism as criticism of the moment at the moment; to Weaver what this world needs in its discourse is the time needed for reflection, arguing that the "constant stream of sensation . . . discourages the pulling-together of events from past time into a whole for contemplation"; the "absence of reflection" that is characteristic of fragmentation leads to our not preserving communal memory (*Ideas* 111).

Why did this work? Weaver thinks that it's because the speakers enjoyed the privilege of having, or taking, the right to assume that precedents are valid and that, in general, "one may build today on what was created yesterday" (*Ethics* 169). "Such presumption . . . instead of being an obstacle to progress, furnishes the ground for progress" (169). The purpose was not so much to make people think as to "remind them of what they already thought" (172).

> *Communication [is] a generalized form of love.*
> —Kenneth Burke (*Attitudes Toward History*)

Like Weaver, Burke wrote when there was the feeling that traditional ways of acting or communicating were collapsing, much like some of us feel today. *Permanence and Change* was written in the early days of the Great Depression, and Burke says he wrote it to help keep himself from falling apart; "orientation" and "interpretation" were his starting points, then "division," or what he calls "perspective by incongruity," then "motivation." The result, he says in his prologue, "is a kind of Transformation-at-one-remove, got by inquiry into the process of transformation itself" (what he says he's writing about, actually, is *communication*) (xlvii). Burke mulls "over the possibility of a *neutralized* language, a language with no clenched fists" and then says he can't help seeing that "Poetry uses to perfection a *weighted* language. Its winged words are weighted words" (*Permanence* liii). Such weighting arises from group relationships, that is, the social, which is grounded in a sociality of cooperation. Group relationships depend on and are characterized by attempts at identification—"at creating a sort of mind-meld between speaker and hearer in which the rhetor *understands* the other's mind, her hopes, fears, and needs" (Anderson 462). Burke differs from Weaver in that to him rhetoric is not rooted so much in any past condition of human society as it is in an "essential function of language itself, a function that is wholly realistic, and is continually born anew; the use of language as a symbolic means of inducing cooperation in beings that by nature respond to symbols" (*Rhetoric* 43). Because we are "in the state of the 'fall,' the communicative disorder that goes with the building of the technological Tower of Babel," we need to foreground cooperation, to build relationships in the public realm (139).

In *Permanence and Change*, Burke says that any effective and "sound" communication medium arises out of some cooperative enterprise. The mind, as largely a linguistic product, "is constructed of the combined cooperative and communicative materials. Let the system of cooperation become impaired, and the communicative equipment is correspondingly impaired, while this impairment of the communicative medium in turn threatens the structure of rationality itself" (163). When he's talking about the communicative norms that match the cooperative ways of a society, he says, "Only those voices from without are effective which can speak in the language of a voice within" (*Rhetoric* 63).

What often is missed about Burke is his addition of *division* to this weighting, which is essential to his concept of cooperation and identification; we create divisions, Burke says in *A Rhetoric of Motives*, because of the human urgency to push away what we perceive to be different from us—and rhetoric can show us that we need these differences that must be always surrounding us because we draw on them for what we cannot draw from our own self. If there were no division— and I think the concept of space works here as well—we might not have the listeners we need in order to achieve cooperation, to gain identification. Who would be the objects of our "pure persuasion" (270-74)?

What I want to suggest is that Burke, like Weaver, calls for recognizing the importance of a dividing space, or the creating of a necessary space, before there can be identification. And identification is necessary for persuasion. Division is necessary to identification; after all, we wouldn't want what Burke calls "ultimate identification": You could kiss your own lips, and have all the fun to yourself" (*Rhetoric* 329).

Weaver's metaphor of the stereopticon creates no division or space, just one image fading in as another fades out. Elsewhere I have used the metaphor of the stereoscope to illustrate what Burke means by change of identity becoming possible through perspectives by incongruity (T. Enos, "'An Eternal Golden Braid'" 108). In order to grow, we must change our perception of a "truth." "Change of identity," Burke says, "is a way of 'seeing from two angles at once'" (*Attitudes* 271): Like the stereoscope that lets us view two images that are nearly the same, but because they are taken from a slightly angular view, each offers a different perspective. The two images project to our consciousness, which unites them into one, three-dimensional view. It is ethos that underlies this process of identification, a generative ethos that "is always in the process of making itself and of liberating hearers to make themselves" (Corder, "Varieties of Ethical Argument" 14).

Burke helps us understand that rhetoric is about constructing and reconstructing, a necessary process before any deconstructing can begin. "Rhetoric is *par excellence* the region of the scramble, of insult and injury, bickering, squabbling, malice and the lie, cloaked malice and the subsidized lie" (*Rhetoric* 19). But rhetoric has benign elements too, ones that include "resources of appeal ranging from sacrificial, evangelical love, through the kinds of persuasion figuring in sexual love, to sheer 'neutral' *communication*" (19).

The aim of this "verbal atom cracking," the ritual we undertake when we enact a discourse, is both to divide—to create space and then to create identification—and achieve identification through coming together in some way (see T. Enos, "'Verbal Atom Cracking'" 65). This is the way we derive our perspective, a perspective of seeing from two angles at once; thus, all perspectives are perspectives by incongruity. Verbal atom-cracking may sound like dissociation only, but Burke says it is also identification because its effect is to blur differences and accentuate some oneness, with oneness meaning a bringing together of various aspects of the most diverse things with a clash of identity; we therefore bring into the process reconstructing as well as constructing and deconstructing.

Like Weaver and Burke, Habermas thinks cooperation is essential for communication, for persuasion. Linking Weaver (who is our most conservative rhetorical figure) and Habermas (who, if not orthodox Marxist, bases much of his theory on the classical Marxist view) seems risky, yet both believe in universal principles and warn against relativism unchecked; both see the values of democracy as being basic to human communication. All three build their theoretical domains on a theory of language, although Habermas moves more toward a conversation model, certainly a link to Corder, as a response to our culture's shifting to an information model. And all three see historical study as a method for challenging and transforming the present as it rediscovers and reinterprets the past. All three are concerned with the moral and the ethical. What we can learn from Habermas and his rationalist theory of communicative action is how the private individual can influence the public sphere, making possible the rediscovery of (a) civil society.

The public sphere that Habermas explores shares some features of the Greek public sphere, a society where freedom was found in public discourse rather than the private realm (most members of the ancient Greek society in any case were excluded from the public sphere). Habermas's theory shares much with the tradition of civic virtue; both reflect the democratization of decision-making processes—action through practical debate. Farrell points out telling similarities between Aristotle and Habermas in that they both "plainly believe that logos and habituation have something important to do with one another" (149). For the Greeks judgment was conveyed by the concept of *phronésis*, or practical wisdom, which is a practical theory accumulated through action around community issues in the space of public life. One key element, however, is reversed. Habermas's "public sphere" is

> defined as the public of private individuals who join in debate of issues bearing on state authority. Unlike the Greek conception, individuals are here understood to be formed primarily in the private realm. . . . The private realm is understood as one of freedom that has to be defended against the domination of the state. (Calhoun 7)

The public sphere, society, Habermas says, is conceived of as the "sphere of private people come together as a public" (*Structural Transformation* 27).

Gerard A. Hauser explains the public sphere as "a discursive space in which individuals and groups associate to discuss matters of mutual interest and, where possible, to reach a common judgment about them. It is the locus of emergence for rhetorically salient meanings" (21). He goes on to call for the return of deliberative discourse:

> The nature of the public sphere is, arguably, the central consideration conditioning the possibility of a participatory public life. In an era when special interests and the state have reduced politics to mass media spectacle, and "audience" has become an economic variable of spectators expected to applaud and purchase, current deliberations over the public sphere advance a critical antidote. By returning the medium of deliberative discourse to political relations, this discussion reasserts the rational function of citizen action as evaluation and judgment. (20-21)

To Habermas scientific-technological knowledge and interests have been overdeveloped in comparison to moral concerns. Devaluing the moral contributes to a nondemocratic polity. But in a fully rational society, each "moral" sphere would be fully developed as rationality would be identified as covering scientific truth as well as moral and aesthetic knowledge (Seidman 24). The development of moral spheres would thus enhance critical reflexivity. And the rationality developed in each sphere would enrich everyday life through the institutionalization of regular interchanges between expert cultures and daily life (see Seidman). To communicate effectively, we say what is true (objectivity), abide by appropriate social conventions (comity), and represent ourselves truthfully, with sincerity (subjectivity). Habermas's "communicative action" is a process of reaching reciprocal understanding; time and space necessarily are built in order for the crucial self-reflexivity to occur.

The criticism against Habermas's theory of communicative action, which would, or should, lie at the heart of democracy, is that his "rationalist model of public discourse leaves him unable to theorize a pluralist public sphere and leads him to negotiate the continuing need for compromise between bitterly divisive and irreconcilable political

positions" (Garnham 360). I would suggest that face-to-face communication and the skills of rhetoric certainly do influence the outcome of any
debate. I would also suggest that Habermas's theory serves as a critique
of postmodernist thinking that too often offers too much cynicism and
relativism. I think we might draw on Habermas, along with Weaver and
Burke, to help prepare for a call for comity because he shows us that
there can be many "publics" (they can be competing, of course), that the
functioning of the public sphere can shift from critical debate to
negotiation, that the foundations of his social theory rest on
communication aimed at reaching agreement, and, most importantly,
that his ideas for communicative action require a self-critical attitude,
some reflective level within a cultural tradition.

CORDER AND THE CALL FOR COMITY

When I read Weaver's observation that "whereas Plato built the
cathedrals of England, Aristotle built the manor houses" (*Ideas* 119), I
could not help but think of the "cathedrals" Weaver, Burke, and
Habermas have built in the sphere of rhetoric—and the "manor house"
(or, perhaps better, "cottage") that Corder spent his whole life designing
and building—but never completing. I want now to focus on Corder to
pinpoint his view of how we might resolve some of the problems that our
society faces, which I've tried to isolate in this chapter. Corder teaches us
how we might communicate in both the private and public spheres, how
we might negotiate mutual agreement in a world that seems to have little
stability, how we might see that rhetoric is indeed a sort of generalized
love, how we might build into our discourses time and space necessary
for reflection—and change our course from incivility to civility.

Building a spacious rhetoric that enabled him to construct a public
self that would be seen as continuous with the private is what occupied
Corder. He wasn't interested in finishing that process or in fully
articulating a new rhetoric. This is why the universe of invention is
paramount in his rhetoric: resistance to closure, striving for open-
endedness. He was intensely interested, however, in articulating certain
ethical codes that can illuminate a modern ethos—indeed, a postmodern
ethos, too, though he seemed ambivalent about the postmodern turn.
Corder's rhetoric is, always, an emerging rhetoric.

To be always exploring persuasive discourse as emerging, as risk-
taking, as a turn toward comity by way of ethos, allowed Corder to draw
on Aristotle's concept of civic virtue, always voicing the "good citizen"
who has internalized through habituation moral values, not unlike, of
course, the learning of manners that are part of the civilizing process.
Perhaps more strongly than any other twentieth-century rhetorician,
Corder anchors for us the lesson given 2,500 years ago—*virtù* gained by
habit through practice, in this case conscious attention to ethos and its
generative capacity that can lead to a consciousness of comity.

What we can learn from Corder, as well as from Weaver, Burke, and Habermas, is how we can build public selves to make our private selves transparent—as the earlier-mentioned Great Exhibition of the nineteenth century did. We need time and space first of all, Corder says:

> We must pile time into argumentative discourse. Earlier, I suggested that in our most grievous and disturbing conflicts, we need time to accept, to understand, to love the other. At crisis points in adversarial relationships, we do not, however, have time, we are already in opposition and confrontation. Since we don't have time, we must rescue time by putting it into our discourses and holding it there, learning to speak and write not argumentative displays and presentations, but arguments full of the anecdotal, personal, and cultural reflections that will make us plain to all others. The world, of course, doesn't want time in its discourses. The world wants the quick memo, the rapid-fire electronic mail service; the world wants speed, efficiency, and economy of motion, all goals that, when reached, have given the world less than it wanted or needed. We must teach the world to want otherwise, to want time for care. ("Argument" 31)

Any kind of specialized language—slang, professional jargon, incivility, any kind of conversational forms that we use without really thinking—stand in the way of communication because they bind space, the specialized language of the speaker prohibiting entry by a hearer into the speaker's discursive universe. Forms of utterance that Corder calls "Pavlov language," forms of reaction, not reflection,

> are the speaker's violation of his or her own space and time, evidence of failure to explore and to know his or her own world. The languages of television and other mass media crowd our time and hurry our responses, to stop our responses entirely. The languages of confrontation set space against space, universe against universe, and cannot wait for meditation time. In such discourse language cannot resonate through time and space. Many instances of failed communication and many instances of communication never attempted are consequences of some violation of space or time or both. ("Studying Rhetoric" 34)

And it is through ethos that we can teach others how to build time for reflection. The Ancients' emphasis on *virtû*, Weaver's on the type of argument, Burke's on identification, Habermas's on a conversational model—all coalesce in Corder's exploration and modeling of ethos. And where in ancient times the public realm took precedence over the private, *civitas* over *civis*, we can see from Weaver to Burke to Habermas and in Corder how the merging of the two allows for an emerging rhetoric, a spacious rhetoric.

An early Corder piece, "Varieties of Ethical Argument" (1978), explores a lifelong quest of his: to try to discover why we listen to some and not others. Corder talks about five possible kinds of ethos: *dramatic* ("literary characters, who are clearly created by their language"), *gratifying* (popular culture figures and politicians, who seem "uniquely fitted to fulfill the need"), *functional* (entertainers and brand names— "serves as a mark of recognition"), *efficient* (a self-completing ethos bound in time, "certain heroes and archetypal figures"), *generative* (an ethos "we need both to hear in others and to make ourselves") (14). A generative ethos, unlike an efficient ethos that is "tied to a particular inventive-structural-stylistic set and completes itself," always "is in the process of making itself and of liberating hearers to make themselves. In this form of ethos, there is always more coming. It is never over, never wholly fenced into the past. It is a speaking out from history into history" (14). Corder gives as one of his examples of generative ethos Weaver's discussion of the Gettysburg Address because Weaver brings in the "transcendent capacity, an ethos torn by the tragedy of a particular time and place, yet awake to the great reaches of time and space beyond the moment and compelled to stewardship of the future" (14). We must learn to "range in space," to live in space "large enough to house contradictions" yet understanding the necessity for change, for seeing from two angles at once; "out of this range grows a command of time, particularly its extension into the future" (14). A generative ethos is a transforming ethos; such an ethos "opens the borders of the discourse to hold extraordinary space and time" (20). Generative ethos can help both speaker and hearer create a world "spacious enough to house [our] antagonists, modifying [our] own world and future in the process"—thus each of us can be a "steward, not an owner, of space and time" (20).

If we can hear only ourselves, we become our own enchained listeners. "We may hear ourselves," Corder goes on, "not another; the other's words may act only as a trigger to release our own, unlocking not the other's meaning, but one we already possessed. When this happens, we are bound in space, caught tightly in our own province" ("Varieties" 20). "Pavlov language," those utterances too often of incivility, stand in the way of communication because they bind space, "prohibiting entry by a hearer into the speaker's discursive universe"; they "are the speaker's violation of his or her own space and time, evidence of failure to explore and to know his or her own world. . . . The languages of confrontation set space against space, universe against universe, and cannot wait for meditation time" (20).

Ten years later Corder laments his earlier conviction about the certainty of ethos residing in the text, of the postmodern rejection of ethos and its study. "Ethos, I've come to learn, is at the very least problematic. I can't have ethos in the text, it seems clear, not on my own previous terms"; furthermore,

> If ethos is not in the text, if the author is not autonomous, I'm
> afraid that I've lost my chance not just for survival hereafter . . .

> but also for identity now. Poststructuralist thought announces
> the death of the author: Language writes us, rather than the
> other way around, and interpretation prevails rather than
> authorship. We have been, whether knowingly or not, whether
> directly or not, part of a 2,500-year-old tradition that allowed
> and encouraged us to believe that ethos is in the text, that
> authors do exist, that they can be in their words and own them
> even in the act of giving them away. Now literary theorists both
> compelling and influential tell us that it is not so, that ethos
> exists if at all only in the perceiving minds of readers, that
> authors, if they exist, do so somewhere else, not in their words,
> which have already been interpreted by their new owners.
> Language is orphaned from its speaker, what we once thought
> was happening has been disrupted. Authors, first distanced,
> now fade away into nothing. Not even ghosts, they are
> projections cast by readers. ("Hunting for *Ethos*" 301)

Jim told me one time that he had spent his whole life trying to write
with a speaking voice. He mentioned his particular interest in why we
listen to some but not others—he wanted to better learn to listen but even
more so to be listened to, to be heard. This is evident in the ethos evident
in any of his writings; what is most striking is that in order to do this, in
order to create the voice of someone the reader wants to listen to, he has
to build time and space into his rhetoric:

> The writer's *living time* has already been radically reduced to
> the writer's *composing time*, and that has been reduced to a
> *presentation time*. Compressing at every stage, the process
> seeks expansion. Across the way, the reader's *living time* has
> been reduced to an *available time*, thence to a *reading time*.
> Where the processes come together, each may be compressed
> into obliteration, but each wants to expand. The compressions
> and expansions that occur here can be cold and unfriendly,
> close and antagonistic. Perhaps instead of confronting each
> other with truths already found, we can learn to accompany
> each other toward what we'll find. ("Academic Jargon" 319;
> emphasis added)

In that space where a reader and a writer meet, there are echoes of things
that not only have already happened but also are inevitable. Where we
might start is with the personal essay, "*a* place to begin to know ourselves
and to show ourselves without confrontation" ("Academic Jargon" 317).
"A perpetual sorrow in human communication," he laments, "is that
often we only announce ourselves to each other, write as authorities
before and at each other, but don't give each other much time" (317). In
that space where a reader and a writer meet, there are echoes of things
that not only have already happened but also are inevitable.
    The humorous little piece, "The Last Essay on Smoking," is an
example of how the personal essay perhaps is the best vehicle to
dramatize how minds can be pretty small by way of an "authoritarian

vision which empowers them to intrude in the lives of others." "It's over [the freedom to smoke], I guess, though a few of us will keep on smoking until the end. We'll be sad, though, remembering that civility cannot survive among inflexible states of mind, remembering, too, that tobacco smoke isn't nearly as dangerous as authoritarian piety" ("Academic Jargon" 323).

To Corder building time and space means that we habitually dwell in the inventive universe and that we always be cognizant of the ethos in our discourse. In nearly all of his writings, Corder worked with ethos, classifying it, defining it, questioning it, hunting for it—and always modeling it. Corder does not seek so much to disprove some charge or even to convince readers otherwise. Corder's ethos, his "argument by indirection" (see, especially, Keith Miller's and Tilly Warnock's chapters in this volume), and thus much of the effectiveness—indeed, the charm and the beauty—of his ethos is that he writes to a reader who agrees with his narrative, and he takes pleasure in the mutual awareness. Corder's own ethos, as I've argued before, is a transforming ethos because it opens the borders of discourse where comity can reside and because of his stylistic choices and representative anecdotes and examples that enable him to engage with the reader in dialogue. "Voice as transforming ethos demands this dialogic action, where both writer and reader [speaker and listener] are aware of, and enjoy, the engagement" (T. Enos, "Voice as Echo of Delivery" 194).

One of the more important things that Weaver, Burke, Habermas, and Corder tell us, and show us, is that we must not lose through time compression the kind of rhetoric that invites reflexivity. Richard Enos suggested to me after reading a draft of this chapter that when Martin Luther King's "I Have a Dream" moved from oral rhetoric to the reflective dimension of writing, it compelled generations long after the actual event to reflect on appropriate action. Through his writings Corder invites us to see that one of the things that is lost in our time—and I've tried to show the elocutionary and Victorian opposites—is the value of time. Time gives us the opportunity for reflection. Because writing stabilizes language, as opposed to the immediate and spontaneous interaction of primary (oral) rhetoric and even secondary (electronically mediated) rhetoric, we can actually create the time by freezing thoughts that invite reflection and that give us the opportunity to not only self-reflect but to reflect as a community.

I am not arguing here for another of those pendulum swings that too often seem characteristic of our field. No society has been, or can be, I think, a shining model of comity. A society that is wholly civil not only is difficult to imagine but perhaps even undesirable. We don't have to argue for a conflict-free society, but we can work toward more constructive, and civil, ways of expressing opposition. I am also offering a "partial" defense of modernism. If exploring ways of suspending urgency, for a moment, can lead to greater comity among us, then this is in itself an opening up to the spaciousness of rhetoric.

WORKS CITED

Anderson, Virginia. "Property Rights: Exclusion as Moral Action in 'The Battle of Texas.'" *College English* 62 (March 2000): 445-72.

Aristotle. *The Nicomachean Ethics*. Trans. H. Rackham. Ed. E. H. Warmington. Cambridge: Harvard UP, 1926.

Auspitz, Katherine. "Civic Virtue: Interested and Disinterested Citizens." *Civility and Citizenship in Liberal Democratic Societies*. Ed. Edward C. Banfield. New York: Paragon, 1992. 17-38.

Banfield, Edward C. "Introductory Note." *Civility and Citizenship in Liberal Democratic Societies*. Ed. Edward C. Banfield. New York: Paragon, 1992. ix-xii.

Booth, Wayne C. *Modern Dogma and the Rhetoric of Assent*. Notre Dame, IN: U of Notre Dame P, 1974.

Boswell: Life of Johnson. Ed. R. W. Chapman. London: Oxford UP, 1970.

Burke, Kenneth. *A Rhetoric of Motives*. Berkeley: U of California P, 1969.

—. *Attitudes toward History*. Los Altos, CA: Hermes, 1937.

—. *Permanence and Change*. 2nd rev. ed. Berkeley: U of California P, 1984.

Calhoun, Craig. "Introduction." *Habermas and the Public Sphere*. Ed. Craig Calhoun. Cambridge, MA: MIT P, 1992. 1-48.

Carbone, Elisa. "Students Behaving Badly in Large Classes." *Promoting Civility: A Teaching Challenge*. Ed. Steven M. Richardson. San Francisco: Jossey-Bass, 1999. 35-43.

Carter, Stephen L. *Civility: Manners, Morals, and the Etiquette of Democracy*. New York: Basic Books, 1998.

Corder, Jim W. "Academic Jargon and Soul-Searching Drivel." *Rhetoric Review* 9 (Spring 1991): 314-26.

—. "Argument as Emergence, Rhetoric as Love." *Rhetoric Review* 4 (Sept. 1985): 16-32.

—. "Hunting for *Ethos* Where They Say it Can't Be Found." *Rhetoric Review* 7 (Spring 1989): 299-316.

—. "Studying Rhetoric and Teaching School." *Rhetoric Review* 1 (Sept. 1982): 4-36.

—. "Varieties of Ethical Argument with Some Account of the Significance of *Ethos* in the Teaching of Composition." *Freshman English News* 6 (Winter 1978): 1-23.

Crowley, Sharon. "The Man from Weaverville." *Rhetoric Review* 20.1/2 (2001): 66-93.

Dowd, Maureen. *Arizona Daily Star*. Feb. 15, 2001: B7.

Eias, Norbert. *The Civilizing Process: The History of Manners and State Formation and Civilization*. Trans. Edmund Jephcott. Oxford: Blackwell, 1994.

Enos, Richard Leo. E-mail to the author. October 2, 2000.

Enos, Richard Leo, and Karen Rossi Schnakenberg. "Cicero Latinizes Hellenic Ethos." *Ethos: New Essays in Rhetorical and Critical*

Theory. Ed. James S. Baumlin and Tita French Baumlin. Dallas: Southern Methodist UP, 1994. 191-210.

Enos, Theresa. "'An Eternal Golden Braid': Rhetor as Audience, Audience as Rhetor." *A Sense of Audience in Written Communication.* Ed. Duane Roen and Gesa Kirsch. Newbury Park, CA: Sage, 1990. 99-114.

—. "'Verbal Atom Cracking': Burke and a Rhetoric of Reading." *Philosophy and Rhetoric* 31.1 (1998): 64-70.

—. "Voice as Echo of Delivery, Ethos as Transforming Process." *Composition in Context.* Ed. W. Ross Winterowd and Vincent Gillespie. Carbondale: Southern Illinois UP, 1994. 180-95.

Erasmus, Desiderius. *De civilitate morum puerilium. Collected Works of Erasmus.* Ed. J. K. Sowards. Trans. and Annot. Brian McGregor. Toronto: U of Toronto P, 1985. 269-89.

Farrell, Thomas B. *Norms of Rhetorical Culture.* New Haven: Yale UP, 1993.

Garnham, Nicholas. "The Media and the Public Sphere." *Civility and Citizenship in Liberal Democratic Societies.* Ed. Edward C. Banfield. New York: Paragon, 1992. 359-76.

Habermas, Jürgen. *The Structural Transformation of the Public Sphere.* Trans. Thomas Burger. Cambridge, MA: Harvard UP, 1989.

—. *The Theory of Communicative Action.* Trans. Thomas McCarthy. Boston: Beacon, 1981.

Hauser, Gerard A. "Civil Society and the Principle of the Public Sphere." *Philosophy and Rhetoric* 31.1 (1998): 20-40.

Irwin, Terence. *Aristotle's Ethics.* New York: Garland, 1995.

Jarratt, Susan C., and Nedra Reynolds. "The Splitting Image: Contemporary Feminisms and the Ethics of *êthos.*" *Ethos: New Essays in Rhetorical and Critical Theory.* Ed. James S. Baumlin and Tita French Baumlin. Dallas: Southern Methodist UP, 1994. 37-64.

Kesler, Charles R. "Civility and Citizenship in the American Founding." *Civility and Citizenship in Liberal Democratic Societies.* Ed. Edward C. Banfield. New York: Paragon, 1992. 57-74.

Norgaard, Rolf. "The Rhetoric of Civility and the Fate of Argument." *Rhetoric, the Polis, and the Global Village.* Ed. C. Jan Swearingen and Dave Pruett. Mahwah, NJ: Erlbaum, 1999. 245-53.

Orwin, Clifford. "Citizenship and Civility as Components of Liberal Democracy." *Civility and Citizenship in Liberal Democratic Societies.* Ed. Edward C. Banfield. New York: Paragon, 1992. 75-94.

St. George, Andrew. *The Descent of Manners.* London: Chatto and Windus, 1993.

Schlesinger, Arthur M. *Learning How to Behave: A Historical Study of American Etiquette Books.* New York: Macmillan, 1946.

Schneider, A. "Insubordination and Intimidation Signal the End of Decorum in Many Classrooms." *The Chronicle of Higher Education* 27 Mar. 1998: A12-A14.

Shils, Edward. "Civility and Civil Society." *Civility and Citizenship in Liberal Democratic Societies.* Ed. Edward C. Banfield. New York: Paragon, 1992. 1-15.

Seidman, Steven. *Jürgen Habermas on Society and Politics: A Reader*. Boston: Beacon, 1989.

Tannen, Deborah. *The Argument Culture*. New York: Random, 1998.

Thucydides. *History of the Peloponnesian War*. Trans. Charles Foster Smith. Cambridge, MA: Harvard UP, 1919.

Trimbur, John. "Agency and the Death of the Author: A Partial Defense of Modernism." *JAC* 20 (Spring 2000): 284-98.

Will, George W. "Coarsening of Society Product of New Economy." *Arizona Daily Star* 19 June 2000: B7.

—. "Teach Virtues Rather Than 'Values.'" *Arizona Daily Star* 25 May 2000: B7.

Weaver, Richard. *The Ethics of Rhetoric*. Davis, CA: Hermagoras, 1985.

—. *Ideas Have Consequences*. Chicago: U of Chicago P, 1948.

9

Toward an Adequate Pedagogy for Rhetorical Argumentation: A Case Study in Invention

Richard E. Young
Professor Emeritus, Carnegie Mellon University

☙Some time ago two colleagues[1] and I completed a three-year research project designed to study the contribution of interactive videodisc technology to the teaching of moral argument. The participants in the first year of the project were students in my upper-level course in rhetorical argumentation at Carnegie Mellon University. The project compared the instructional value of videodisc technology with that of film and written text. The videodisc has some of the vividness and immediacy of film, but unlike film it offers students the ability to interact by typing responses to questions posed on the computer screen. The particular videodisc we worked with, which won the 1989 EDUCOM/ NCRIPTAL Award for Best Humanities Software, dealt with Dax Cowart, [2] whose life epitomizes many of the central problems in the growing controversy over euthanasia. Cowart, who was what we often think of as the All-American Boy (handsome, football player, rodeo rider, jet pilot), was burned over his entire body in a propane explosion and was forced to undergo more than a year of incredibly painful treatments, in spite of his protests and repeated requests for help in killing himself. The treatments saved his life but left him blind, crippled, and severely scarred.

The original purpose of the research project was to determine whether the interactive videodisc had a greater potential for helping students develop considered judgments on some issue in the Dax Cowart case than did written text on the case or a film that had been made of the case for use in medical ethics courses. The present study is an offshoot of the original project, an attempt to understand one of the unanticipated results.

The course had several routine objectives (improved writing skills, a better understanding of the nature of argument, etc.), but the principal objective was to help the students develop a considered judgment on an issue in the euthanasia controversy and to argue that judgment to a particular audience. This is not just one more objective for an advanced course in rhetorical argument; it is a fundamental objective of rhetorical

159

education. One might argue that it is fundamental to liberal education as well.

The interactive videodisc we used has two characteristics useful in achieving this objective: (a) It provides the student with a rich database on the Dax Cowart case and (b) in presenting the various positions of the persons involved in the controversy, it offers, in effect, alternative arguments on the question of whether Dax Cowart's request to discontinue his treatments should have been granted, arguments against which students could test their own. It is designed to help students engage in what John Dewey called a "reflective experience" (150), which entails coming to know in a situation that is initially perplexing, a movement from questioning and tentative judgments to their careful consideration in light of the available evidence and finally to more refined judgments that have been tested against the evidence, that is, to considered judgments.

We anticipated that the interactive video would engage the students intellectually and emotionally more than the other mediums—that the vividness of the visual images, the multiple paths it offered through the data, and particularly the intractability of competing claims of the participants would confront the students more fully with the complexity of the case and the poignancy of the dilemmas faced by each of the participants. We hoped students using the videodisc would move beyond the kind of shallow understanding of issues that teachers in argumentation courses must often settle for. To determine whether students were actually developing considered judgments, various kinds of data were collected from each of the students: videotaped think-aloud protocols as they worked with the interactive videodisc (which also recorded their typed responses), arguments written for the class, pre- and posttests designed to measure changes in attitude, and interviews after the course had ended.

I tried to help my students develop skill in argumentation and reach the course objective in a more or less conventional way. I introduced them to Toulmin's logic, to the classical art of arrangement, and to various theoretical concepts in rhetoric, such as stasis and rhetorical situation—all of which served as analytical tools for criticizing their own work, the work of the other students, and the arguments in the text. The text for the course was Annette Rottenberg's *Elements of Argument*, a solid introduction to argumentation that included a discussion of Toulmin's logic and a collection of articles that presented conflicting views on euthanasia. During the course students wrote and reworked four papers in which they were asked to develop arguments that presented their own judgments on some aspect of the euthanasia controversy that they saw as interesting and problematic. The Dax Cowart case and the videodisc were introduced about halfway through the course as an instantiation of the euthanasia controversy studied in more general terms earlier.

Unfortunately, the results of the project gave us good reason to believe that the course did not achieve its principal objective: That is, it did not appear to help students develop considered judgments on ethical problems. The quantitative results for every student indicated that no statistically significant change occurred in their thinking relative to this objective. This was surprising since most of us, I think, would predict that the thinking of at least a few students would have changed to some significant degree. But, to my dismay, their pretests were the best predictors of their posttests.

Even though the pre- and posttest arguments written for the course did show other kinds of change, these were nearly as distressing as the statistical results, for they suggested that the students had simply become more expert at arguing their initial positions. Apparently, I had taught them to be more articulate about their positions, to explain them better, to perhaps be more persuasive about them, and to be more effective in their criticisms of other positions. I see now that they had learned quite well what I had actually taught, which was not without value; however, it wasn't what I was trying to teach, what I thought I was teaching. That is, I had not taught them to be more critical of their own positions and more sensitive to the complexities of the controversy. If one judges the success of a course solely on whether it achieves its primary objective, this course was a failure. The research project, with its careful effort to capture and examine the outcomes of instruction, helped me to see what I might otherwise have ignored or explained away; it also raised some important questions about rhetorical invention and the teaching of argument.

I would like to look carefully at the work of a student I will call "Susan," focusing in particular on the extraordinary stability in her thinking throughout the course. In 1990 she was a senior majoring in English; she was not an outstanding student, but she was serious and competent, in many ways typical of the others in the course. She begins her work on the interactive videodisc with the computer asking whether she thought "Dax should be allowed his request to discontinue his treatment," the principal question posed by the Dax Cowart case. She immediately answers "yes" (4:23:39),[3] and for the next hour and twenty minutes of work with the videodisc, she continues answering an unqualified "yes" despite being confronted with compelling counter-arguments and facts inconsistent with her position.

The claim in her primary argument is that Dax should be allowed to commit suicide because, she argues, we all have a right to dispose of ourselves as we choose, assuming that we are competent (that is, that we are in full possession of our faculties and understand the situation), and Dax is clearly competent to make the choice. Backing for the warrant seems to be the assumption that we own ourselves unconditionally. In her responses to questions posed by the computer, she appears to waffle about why suicide would be a reasonable choice to make: She says at various times that he is in extraordinary pain; that he will be blind, crippled, and scarred; that he has little chance of recovering; that he won't be able to do the things that make his life worth living.

Finally, she argues that those caring for Dax should have helped him commit suicide because he was too crippled to kill himself and that they had a responsibility to help, an assertion she never supports. For example, she types that "the nurse seems to want to do what is right but the slim chance that she will have guilt feelings is outweighing the pain of Dax's. this wrong. she should, if she is unable to terminate his life, find someone else who will do it. this is her responsibility" (4:35:35).

When confronted with arguments that conflict with her own, she simply rejects or ignores them. For example, when the doctors argue that Dax may not be completely blind after his treatment and may have at least some function restored to his hands, she says out loud, "That's just rationalizing. I would still let him make the decision" (5:1:44). Later while being interviewed, she remarks that "the doctors sort of didn't make much of an impact [on me]. I mean they did, but, I don't know, I guess once I made my decision I didn't think there was anything . . . could've changed my mind." As Jim Corder observes, "we may hear ourselves, not another; the other's words may act only as a trigger to release our own, unlocking not the other's meaning, but one we already possessed. When this happens, we are bound in space, caught tightly in our own province" ("Varieties of Ethical Argument" 6).

Her position is consistent across the various kinds of data we collected. For example, in the pre- and posttest attitude questionnaires, she is asked to imagine that she is a doctor with a patient who must have treatment in order to live but who has requested that the treatment be discontinued. She is then asked to construct "a situation in which [she] would have no moral or ethical hesitation in granting the request." In the pretest she responds, "If the patient had AIDS and was going to die soon and just wanted to get it over with the least amount of suffering." To the same instructions in the posttest, she responds, "In the Dax Cowart case I would have granted it," overlooking the difference between the two cases, that is, that there was good reason to believe that Dax was not going to die soon if he received treatment.

And the concluding paragraphs of her pre- and posttest papers show a similar consistency. Her argument for assisted suicide rests on the belief that the individual in a sense owns his own person and, if competent to make a judgment, has a right to choose what is to be done with it. Her pretest paper concludes with her saying that "If Terence Cardinal Cooke [whose argument against euthanasia she had read] is ever faced with a situation in which a loved one is suffering, he will choose what 'he admits to be right,' and we, in turn, will choose, but it is essential that we be allowed to have the choice." And her posttest: "If we are to be human and uphold the idea that people are entitled to their own opinions, then we have to believe Dax."

The final scene on the videodisc shows Dax, now a practicing lawyer, playing with his dog and talking about his recent marriage. Upon seeing him, she bursts out, "Is this him now? Yes. Jesus! This isn't fair. That's great. But unfortunately you can't make these kinds of decisions. . . . You can't predict the future. You have to make the decision before hand" (5:38:39). In the interview at the end of the study, she remarks about her work with the videodisc that "I didn't know what . . . the outcome was going to be, I hadn't heard of the case at all, so I was really surprised at the end. It really threw me to see how it worked out." She overlooks the fact that she had been repeatedly confronted with predictions by Dax's doctors and his lawyer that he would live and could have a reasonably satisfying life.

How are we to understand her performance? And what implications does it have for rhetorical theory and the teaching of argument? "I think we have not fully considered," writes Jim Corder, "what happens in argument when the arguers are steadfast" ("Argument as Emergence, Rhetoric as Love" 21).

Clearly, her performance has some strengths. Although her arguments are not as strong as one would like, she does have clearly stated and tenable claims, modest strengths but ones that might be developed into something more substantial. Anyone who has ever taught undergraduate argumentation is not likely to denigrate their importance. But given the primary objective of the course, that is, developing a considered judgment on the issue, two weaknesses in her performance are especially notable: Her claims are bare statements of principle, untempered by exception and qualification; and, related to this first weakness, the process by which she arrived at the claims included no effort to test them against the available information, which might have produced more qualified claims or even different claims altogether. The process, as apparent in the think-aloud protocols, presents a position but little evidence of reflective thinking on the question of whether Cowart's demand to die should have been granted.

We can explain these weaknesses in several quite different ways and find some support for each in the data collected in the project: (a) The weaknesses might be the result of immaturity and lack of experience in dealing with real-world problems. Or (b), they might be the result of defensiveness, of being unable to consider alternatives that threaten her own position and to some extent, perhaps, her identity. Or (c), they might be the result of not having a rhetoric that would encourage her to place the abstract and rigid principles from which she argues in some reciprocal relationship with the particularities of the case. Or (d), they might be the result of the course design (for example, too little time to reflect on the case). These explanations may not exhaust the possibilities; and they are not necessarily exclusive, because more than one cause could have influenced the outcome. But they offer us a place to begin an inquiry.

Because all the students in the class were at least twenty years old and because not a single one of them reached the principal goal of the course, developmental and psychological explanations seem less likely than the third and fourth explanations. I would like to argue here that Susan lacked an adequate rhetorical theory, one that would have encouraged her to test her arguments against the facts of the case and the arguments of others, and in the process develop a more carefully considered position. Because the last explanation, course design, may be linked to the former, I want to defer consideration of it for the moment.

One plausible interpretation of what happened is that the students were being taught how to argue more persuasively the positions they brought to the controversy, when what I was supposed to be teaching them, what I thought I was teaching them, was how to both discover and argue a considered judgment about the controversy. The distinction is similar to one Perelman makes between two forms of dialogue: debate and discussion. In a debate, Perelman says, "each interlocutor advances only arguments favorable to his own thesis, and his sole concern with arguments unfavorable to his is for the purpose of refuting them or limiting their impact." In a discussion, on the other hand, "the interlocutors search honestly and without bias for the best solution to a controversial problem" (37-38). The distinction is useful, even though in practice it is difficult to maintain because in an actual dialogue, one intention at a given moment may give way to another, or they may blur into each other.

Whether we enter a dialogue with the intent to debate or to inquire by means of discussion seems to depend on whether we think our position is true or whether we have doubts about its adequacy. To put it another way, how we enter the dialogue appears to depend heavily on the epistemological status of our knowledge, what, that is, we believe its status to be. When we are convinced we have the truth, participation in a dialogue is likely to be treated as an opportunity to persuade others that we have it. Or it may be treated as an opportunity to increase the strength of our position by learning what the opposing arguments are. For those who believe they have the truth, problems of rhetoric become largely problems of presentation. And the role of rhetorical invention becomes the relatively modest one of helping us find ways of arguing the preexisting claim in a persuasive way.

At times, however, we are not sure about the adequacy of our knowledge, in which case a tentative claim becomes in effect a hypothesis that requires further inquiry and some sort of testing against the available evidence. At other times we may find ourselves immersed in a felt difficulty, and, genuinely perplexed, we shy away from even the tentative claims that may spring to mind; we have questions but no answers. In both situations the rhetorical process may entail not just knowledge-presenting but knowledge-making as well. Here it is not conviction that dictates the character of the rhetorical activity but the need to know (as, for example, in the deliberations of careful jurors in a courtroom trial or of academic faculty trying to decide fairly a complex tenure case). John Gage describes this kind of activity when he remarks that

> knowledge can be considered as something that people *do*
> together, rather than as something which any one person,
> outside of discourse, *has.* Knowledge can be said to be valid,
> that is, to the extent that it can be shared, and is likely to need
> modification when minds bring new understanding to things
> thus known. Rhetoric can be viewed as dialectical, then, when
> knowledge is seen as an *activity,* carried out in relation to the
> intentions and reasons of others and necessarily relative to the
> capacities and limits of human discourse, rather than a
> *commodity* which is contained in one mind and transferred to
> another. ("An Adequate Epistemology for Composition" 156)

If one has the truth, the tendency is to try to prevail over other positions; if one has doubt, other positions may be a means to new knowledge.

For the person who is convinced he knows, rhetorical invention has one sort of character. For the person who does not know but seeks to know by engaging others in an inquiry, invention has another character, even if, oddly enough, both persons may be working, on the face of it, with the same art of invention. We can see an instance of this curious duality of purpose of the rhetorical art in, for example, two traditional definitions of the enthymeme. The enthymeme has been frequently defined as a truncated syllogism (Schnakenberg 153-61). But it is also, less frequently, defined as probabilistic argument, a kind of argument based not on self-evidence or incontrovertible truth but on the shared experiences of a community and created, in part at least, by dialogic exchanges with other members of the community (Gage, "Enthymeme" 223). The difference is important, because in the light of the former definition, the enthymeme tends to become a formal principle of argument, a syllogism with a missing proposition. With the latter definition, the enthymeme becomes, as Gage remarks, "a sort of metonymy for the whole rhetorical activity of discovering the basis for mutual judgment" (157).

For Susan, who believes she knows the truth, the enthymeme is only a formal principle, not an argument emerging from a process of discovery—a kind of activity Susan, in fact, does her best to avoid. She cheers comments that confirm her own position: For example, when Dax is explaining his right to choose, Susan says out loud, "Right . . . you're right. You are so right. I completely agree" (5:11:31). And she argues constantly with the voices on the videodisc: For example, when the nurse says she couldn't bring herself to help Dax commit suicide, Susan says, "You should have." And when it is pointed out that Dax may not be entirely blind and may recover some use of his hands, she says, "That's just rationalizing" (5:0:50). Certitude tends to induce a predisposition to look no further, confirming the mind in its habitual grooves and encouraging it to ignore whatever might threaten to dislodge it. Certitude, then, subverts dialogue-as-discussion and encourages debate. Similarly, it tends to subvert the value of the interactive videodisc, since the videodisc is essentially a tool designed to stimulate discussion and inquiry. Whether we believe we know or whether we doubt and question determines to a large extent what we understand the nature of rhetorical invention to be and the use we make of it as we create our arguments.

Susan and the other students showed me that I was not working with a conception of argumentation appropriate to my educational objective; furthermore, it was inappropriate to the educational technology I was using. I clearly needed better ways of moving students beyond easy and rigid positions, of cultivating a state of mind characterized by doubt and questioning rather than certitude and assertion. What was needed is not likely to be found in what seem to me to be the major trends in argumentation pedagogy since the 1950s—that is, developing ways to improve the formal features of student arguments (in helping them articulate a thesis, support their claims, structure the parts of the argument, etc.) and developing more effective and appropriate approaches to reasoning (for example, Toulmin's *Uses of Argument*)—as valuable as both these trends have been. It is also clear that technological innovations like the interactive videodisc must be buttressed by appropriate rhetorical theory if they are to be effective.

More specifically, if we are to develop effective courses in rhetorical argumentation that have reflective experience as a central goal, two changes seem to me to be essential. First, we need an art of rhetoric that encourages inquiry and discussion; the present art, the one I relied on in the course, while valuable for many purposes, does not do this well. It emphasizes abstract principle, the self-evident, and the axiomatic in contrast to the particular, the practical, and the probabilistic; and it emphasizes technical knowledge and craft in contrast to effective action. It tends to devalue shared experience, communal belief, the situational and contingent, and the prudential. The dominant rhetorical tradition and our textbooks on argumentation, which are informed by this tradition, have been and still are essentially eristic. In it, rhetorical action is better exemplified by formal academic debate or legal argument in a court of law than it is by, say, jury deliberation, or labor mediation, or the anguished arguments between Dax and those who sought to help him. Because our students bring this tradition with them into the classroom, any alternative to it needs to be taught explicitly; we cannot assume that it too is part of their rhetorical inheritance even though it is part of rhetorical history.[4] The need for explicit instruction in another sort of rhetoric is also apparent when we consider our natural tendency to prefer our own positions and to fault the positions of others. There is little in the dominant tradition that provides a counterpressure to this tendency.

Second, because certitude tends to transform discussion into debate, I think we need to build into rhetorical invention what for want of a better term might be called a *problemology*. We need to find ways of moving students beyond easy notions and into what Dewey described as the "twilight zone of inquiry" (148-49), that is, a state of mind where questioning rather than asserting and bolstering assertion characterizes thinking. Such a state strikes me as fundamentally important for the instructional goal I had set for myself, for it is the motive for reflective thought. Without the experience of a problematic situation and without explicit instruction in an alternative rhetoric that promotes inquiry, it seems unlikely that students will abandon their natural inclinations and their rhetorical inheritance.

The history of rhetoric and dialectic can provide us with help, though until recently we have not sought it there. The long tradition in rhetoric that emphasizes practical wisdom (that is, *phronésis*) and the heuristic function of languaging still lives, but in the shadow of a more familiar tradition. I have in mind the relatively neglected contributions of the Sophists, with their concept of *dissoi logoi* (Enos, *Greek Rhetoric before Aristotle* 57-90; Sloane); Cicero's *De Oratore*, with its emphasis on practice and arguing both sides of the case (Enos, *The Literate Mode of Cicero's Legal Rhetoric* 33-45; Sloane); sixteenth-century casuistry (Jonsen and Toulmin), with its emphasis on the particularities of rhetorical situations, paradigm cases, and analogical thinking; Abelard's *Sic et Non* (Norton), a way, he says, to "stimulate tender readers to the utmost effort in seeking the truth" (20), that is, a formal method for creating doubt in the minds of students.

More recent work might also be helpful: for example, Dewey's notion of reflective experience (139-51), problem formulation in tagmemic rhetoric (Young; Young, Becker, and Pike 71-98), Rogerian argument (Teich; Young, Becker, and Pike 273-89), Gage's work on epistemology mentioned earlier—all might be useful in developing both an alternative rhetoric and a more adequate pedagogy of argumentation. Jim Corder's work on ethos (for example, "Varieties of Ethical Argument," "Argument as Emergence," "When") also seems to me devoted to the sort of theoretical and pedagogical needs I am talking about. In his writing he continually addressed the intellectual and moral perils of certitude and the isolation of opposition, calling for a rhetoric that is closer to our lives as we live them in the midst of others.

At the end of Susan's protocol, when Dax appears on the computer screen playing with his dog in his home, she remarked indignantly, "Is this him now? Yes. Jesus! This isn't fair." What is interesting about the outburst is that for a moment her conviction appears to have been shaken. If I understand her statement, she is not saying that the final scene on the videodisc shows that Dax had been unfairly treated but that she herself has been unfairly treated by being confronted with facts that tend to subvert her position. Having been so adamant about helping Dax end his life, she is distressed when she learns the extent of his recovery. I would argue that her statement marks the entry of doubt into her thinking and the awareness of other possibilities. I would also argue that it marks a point at which dialogue-as-debate might be transformed into dialogue-as-discussion. Unfortunately, it was only at the end of her work with the videodisc that she might have been ready to use what it can offer and to begin moving toward a more considered judgment.

NOTES

[1]Preston Covey of the Philosophy Department, Carnegie Mellon University, and Robert Kozma, then of the School of Education, The University of Michigan.

[2]An extensive collection of articles and a bibliography on the Dax Cowart case can be found in Kliever, *Dax's Case: Essays in Medical Ethics and Human Meaning.*

[3]The numbers in parentheses after statements by Susan refer to locations where the quoted passage appears on the videotape or on the videodisc printout. Photocopies of these and other materials referred to can be obtained by e-mailing me at ry0e@andrew.cmu.edu.

[4]Jonsen and Toulmin provide an illuminating historical discussion of alternative traditions in *The Abuse of Casuistry.*

WORKS CITED

Corder, Jim W. "Argument as Emergence, Rhetoric as Love." *Rhetoric Review* 4 (Sept. 1985): 16-32.

—. "Varieties of Ethical Argument, With Some Account of the Significance of *Ethos* in the Teaching of Composition." *Freshman English News* 6.3 (1978): 1-23.

—. "When (Do I/Shall I/May I/Is It Appropriate for me to) (Say No to/Deny/Resist/Repudiate/Attack/Alter)   Any (Poem/Poet/Other/Piece of the World) for My Sake?" *Rhetoric Society Quarterly* 18 (1988): 49-68.

Dewey, John. *Democracy and Education.* New York: Free P, 1944.

Enos, Richard Leo. *Greek Rhetoric Before Aristotle.* Prospect Heights, IL: Waveland P, 1993.

—. *The Literate Mode of Cicero's Legal Rhetoric.* Carbondale: Southern Illinois UP, 1988.

Gage, John T. "An Adequate Epistemology for Composition: Classical and Modern Perspectives." *Essays on Classical Rhetoric and Modern Discourse.* Ed. Robert J. Connors, Lisa S. Ede, and Andrea A. Lunsford. Carbondale: Southern Illinois UP, 1984. 152-69.

—. "Enthymeme." *Encyclopedia of Rhetoric and Composition.* Ed. Theresa Enos. New York: Garland, 1996. 223-25.

Jonsen, Albert R., and Stephen Toulmin. *The Abuse of Casuistry: A History of Moral Reasoning.* Berkeley: U of California P, 1988.

Kliever, Lonnie D., ed. *Dax's Case: Essays in Medical Ethics and Human Meaning.* Dallas: Southern Methodist UP, 1989.

Norton, Arthur O. "The Rise of Medieval Universities." *Readings in the History of Education: Medieval Universities.* Cambridge, MA: Harvard UP, 1909. 13-106.

Perelman, Chaïm, and L. Olbrechts-Tyteca. *The New Rhetoric: A Treatise on Argumentation.* Notre Dame: U of Notre Dame P, 1969.

Rottenberg, Annette T. *Elements of Argument.* New York: St. Martin's, 1988.

Schnakenberg, Karen Rossi. "Classical Rhetoric in American Writing Textbooks, 1950-1965." *Inventing a Discipline: Rhetoric Scholarship in Honor of Richard E. Young.* Ed. Maureen Daly Goggin. Urbana, IL: NCTE, 2000. 146-72.

Sloane, Thomas O. "Reinventing *inventio.*" *College English* 51 (1989): 461-73.

Teich, Nathaniel, ed. *Rogerian Perspectives: Collaborative Rhetoric for Oral and Written Communication.* Norwood, NJ: Ablex, 1992.

Toulmin, Stephen Edelston. *The Uses of Argument.* Cambridge, MA: Cambridge UP, 1969.

Young, Richard. "Problems and the Composing Process." *Writing: Process, Development and Communication.* Ed. Carl H. Frederiksen and Joseph H. Dominic. Vol. 2. *Writing: The Nature, Development, and Teaching of Written Composition.* Hillsdale, NJ: Erlbaum, 1981. 59-66.

Young, Richard E., Alton L. Becker, and Kenneth L. Pike. *Rhetoric: Discovery and Change.* New York: Harcourt, 1970.

10
Rhetoric and Conflict Resolution

Richard Lloyd-Jones
Professor Emeritus, University of Iowa

❧Donald Bryant, who wrote extensively on Edmund Burke and taught classical rhetoric at Iowa, once asked me how I defined rhetoric. I knew he was bothered by an academic fashion for including most discourse under the term, and he suspected me of being a shameless eclectic, so I stipulated that it was the art of finding the best available means of persuasion. At that moment, being caught up in the chores of administration, I didn't have the energy to sustain a lexical exploration of the elusive term through a conflict of differing views. I did not resolve a potential conflict; I evaded it. Years later I am still struggling with the question.

Bryant recognized an evasion when he heard it and dropped the topic. Perhaps he was trying to seek my mind, but then he might have simply been proposing a word combat after the manner of Renaissance Wits. I knew that if I had to match backgrounds with him, I would make a fool of myself, so I had chosen to stipulate a traditional definition. Stipulation is useful as an introduction to a legal contract or to legislative enactments precisely because it forestalls discussion. In a sense it defines conflict out of existence. A stipulator wants to narrow the focus. Management often depends on stipulation to keep workers on task. A pragmatic person may choose such a system of definition simply because it is useful in carrying out the day-to-day work of a society by getting on with the work at hand. Bryant probably decided that I was not worth fighting with, that he would not get from me the intellectual rewards of resolving ideological conflict.

For many centuries and most people, the philosophical implications of stipulation are not important. They are for us in this context. The word *stipulation* implies an act of will in observing the world that does not presume or even desire access to absolute truth. One chooses to represent an aspect of reality convenient for meeting some present need. But many people don't separate the word or sentence or discourse from what is represented. They think that if you have the name, you have the thing itself. Among some primitive peoples, having the name was enough to command power over the object or person. That is one of the roots of black magic. For them stipulation is merely stating the Truth.

Indeed, one of the basic assumptions of classical and medieval learning is that grammar and rhetoric are basic studies because they provide a person with direct access to what is real. For Plato reality is in the Idea, the abstraction underlying what the senses recognize, the Word, the *logos* as it appears in John's gospel. The classic Aristotelian definition by genus and differentia presumes a world that is essentially classifiable and thus rationally ordered in the forms of language. The eighteenth-century bishops who described English imagined that an ideal language preceded the dust-up at Babel, and that English should be cultivated to come closer to some Edenic perfection. Correctness in an odd way provided access to godliness and Truth. By contrast, the General Semanticists of the twentieth century in their favorite metaphor of language as a map have tried to emphasize the arbitrariness of language in ordering existence. So have many others in the postmodern world, but for traditional people the map *is* the reality.

This traditional view of language is comforting to kings and tyrants, to bishops and lecture-prone professors, because it supports the idea of hierarchies and resists notions of change and local variation. It enables some teachers to honor mechanical correctness, which is definite, at the expense of elaborated discourse, which is problematic. Thus it prevails in daily discourse because, like stipulation, it forestalls speculation and squishy open-mindedness, which lead to conflict. For such people conflict is bad, and it occurs when some other person doesn't understand the Truth; in extreme situations such other people must be beaten into belief. That justifies war.

One can find examples of contrary views at almost any time wherever there is the acceptance of equivocation. At the practical level, it is implicit even in the classical notion of persuasion. In situations where one cannot know the Truth (or where, perhaps, it doesn't exist), one must discover a means of convincing other people that one's own view is more likely correct. Truth simply can be demonstrated, but probability must be shown persuasively. Still the historical linguistics of the nineteenth century, the biological challenges to the Great Chain of Being, the study of variant social structures of aboriginal peoples, the rise of middle-class democracy, and the general willingness to assert the primacy of individuals above the state or society created a climate that encourages a view of the domains of language in terms of Custom and Convenience rather than Truth and Stability. The Romantic ego (and ultimately Protestant faith) sought a different way to use language as a tool, one that favored individual insight as opposed to absolute rule. Thus the practical affirmation of conflict is part of our basic worldview.

Conflict represents diversity of point of view and desires. Contrasting visions create energy and what the nineteenth century viewed as progress. Conflict is good in that it allows Natural Selection to improve life. Perhaps such views are sound, but conflict also leads to wars, general destruction, shattered families, and social dysfunction. One cannot imagine drama or literary fiction without conflict, *agon*, to engage human attention and identification. It is the justification of sport. Conflict we shall always have with us, but we also believe that blessed are the peacemakers. The challenge to civilized people is to contextualize conflict so that its energies can be directed toward positive ends, and that is where a broader definition of rhetoric can be helpful. Many of the practical strategies of rhetoricians have been rediscovered independently by conflict mediators in the last half century and then massaged to fit new contexts.

Conflict resolution has emerged as a separate field of theoretical study and practical training, but it draws from many long-established disciplines. The collegiate degree programs and research centers dealing with conflict resolution tend to be multidisciplinary and variously constructed. The pragmatic efforts to spread peacemaking skills seem even more far-ranging. Just as some people think that anyone who can write can teach writing, so many people assume that anyone with good intentions can generate agreement. Indeed, some writers are good teachers, and some decent people are natural peacemakers. Still, it is useful to document effective techniques and develop explanations of why they work.

The most visible programs of peacemaking are international. We have been told that the misguided peace of World War I created World War II by fostering resentment among the losers. After 1945 the United States and Allies punished individuals who led the war but helped regenerate the nations they defeated. Possibly, the motives were partly to create economic markets for our goods, but they surely included a premise that people who envisioned peaceful ways of becoming prosperous did not choose violent conflict. Somehow the healthy competition described by Adam Smith for an agrarian economy was to become a model for world competition. That is, we redefined the objective of a peace treaty from vengeance and reparations to restoration in seeking a common good. The Marshall Plan and the parallel rebuilding of Japan did not abolish war, but they invited one to see peace in terms of the resumption of an orderly pursuit of general well-being. The contexts of war were redefined. Total war swept up innocent and guilty alike in destruction, and total peace worked toward the regeneration of survivors into a sharing community.

The last half of the twentieth century offers numerous examples of the successes and failures of the strategy. For all of that time, Korea remained divided, and no efforts at resolution went anywhere until 2000. In Vietnam we saw forces from the North drive out colonialists, but the winners were in no condition to be generous. In a quarter century, Vietnam has gradually opened negotiations with other nations, and the accumulations of minor agreements can be seen as a means of building trust and resolving the conflict. Despite almost continuous efforts in the Middle East, resolution has proved elusive although there have been a number of promising moves. The Carter mediations were certainly a step toward some kind of accommodation, but terrorists of several stripes have demonstrated that a tiny minority with explosive power can prolong conflict and perhaps deny resolution sought by the great majority. That is, zealots and madmen do not respond well to rhetoric or negotiation.

Still, one cannot refuse to talk because an obdurate few refuse to participate. If rhetoric cannot solve all problems, it can solve or alleviate some. In Ireland, central America, parts of Africa, and elsewhere, violence has been reduced, or delayed, so that forces of reason can have time to operate. The UN, private groups such as the Carter Center, and individuals such as Senator George Mitchell have been brokers on the international scene with some success in turning destruction into positive forces. Each has been reviled for being soft or meddlesome or indecisive or amoral in dealing with criminals, but each has helped nations or subnations get along with each other.

Efforts at conflict resolution among nations have been the most spectacular version of the peacemaking systems because so much is at stake in a nuclear world. As the world's population keeps doubling and technical power multiplies unpleasant consequences of misbehavior, we have to pay attention, but it is easy to become impatient and frustrated because agreement is likely to be slow and requires considerable readjustment of opinions among many people. Even though the number of wars in progress is distressing, contemplation of the number of catastrophic explosions that haven't happened is encouraging.

For most people interested in rhetoric, the techniques of conflict resolution are better examined at the more modest level of daily local affairs. Small disputes are complicated enough to demonstrate the processes; great disputes are overburdened by the slag of history and chauvinism.

Little disputes turn up on the playground or in small-claims court, on the suburban lot lines or in the back-office cubicles. Conflict is part of living in restricted space. Although international peacemaking deserves headlines, it is at the day-to-day level that the rhetoric of peacemaking has become a grass-roots concern. Where neighborhoods once really included neighbors who somehow knew each other and worked out their problems, now the residents have their friends and workplace somewhere else and barely recognize the person next door. If they have problems, they sue. That is, in both law courts and arbitration systems, one substitutes force for real agreement.

The rhetoric of power and force is hardly new. Gorgias likened it to the coercive effect of drugs, changing the secretions of the body to alter life, a physical force. Helen of Troy, he thinks, was either directed by the gods and so was blameless, or was taken by force and so was blameless. The force might have been that of physical strength, as in rape, or of words, rhetoric. Rhetoric in this case may have been the force of mere beguiling phrases, but that is always suspect. In our day we note that the beguiling appeals of the boss to an employee still imply force. A person of status or heroic stature always displays implications of force when dealing with dependents. One of the objectives of establishing an ethical proof is to create a narrator with some kind of force, an authority that reduces the need for other judgments. We prefer to emphasize ethical proof as establishing trust, which in turn creates dependence. Father knows best, after all. Medieval rhetoric mostly presumed that truth was knowable, scriptural, and that rhetors were simply making that truth available to the unenlightened. The clergy knew the Truth. Hiding behind such an assumption is the force of absolutism.

In the United States, we have institutionalized conflict within the political process. Periodically we have elections that are reported in the press in much the way used to report sports events. Who is ahead, what strategy is being used, what deception is effective, what big guns can be marshaled by the candidate to force—entice—the popular vote? When one is presented a calmly reasoned argument with actual evidence, the press is rarely interested. That is, the rhetoric of problem-solving is not considered attractive enough to draw an audience and justify the fees for advertisers. If this is not actually sport, it is drama with the *agon* to rouse passions. At the moment that one candidate is declared elected, partisans are already planning the next campaign; the conflict is deferred, not resolved. So long as the process is orderly—riot free—the public is satisfied that good governance is ensured. We congratulate ourselves that we all accept the verdicts without violence and bloodshed.

To some extent the partisan contests are muted in legislatures even though party discipline is firm in selecting officers and committee chairs for any legislature. It is not merely that the spoils belong to the victor. Power to act resides in the majority, subject to minor checks in our judiciary and our claim that the essence of democracy is protecting the rights of minorities, and practicalities of daily life chasten the enthusiasm of the executive branch. The glory of our checks and balances is that our elected officials are often encouraged—even forced—to adopt the rhetoric of resolving problems rather than simply using force. I do not pretend that autocratic politics are not attractive to those in power or even that they have no place in certain crises. Some legislatures at some times are extremely partisan. The so-called Gingrich power in our Congress set off such a period. After all, by disposition some people are quarrelers, and sometimes the populace likes stalemates. Indeed, some people suggest that financial markets prefer political stalemates because they lead to predictable situations for investors. That is, many people like conflict for its own sake. For them, a lively conflict improves the quality of life, and resolution merely takes away the fun.

I viewed this kind of relationship secondhand while my wife spent sixteen years in our legislature, eight years in each house. Part of the time her party was in majority; part of the time it was in the minority. For all of that time the governor was of the other party, but he often had to assert an executive point of view that separated him from the legislative leaders of his own party. And of course all of the players had views and values of their own that were not necessarily congruent with the official positions of parties.

As an outsider I was conscious that legislators of whatever party often shared outlooks about the political process that separated them from the public. Circumstances made them into a community that treasured its customs. The Senate, a smaller body with longer terms, was especially aware of its status separate from the world of voters. Both houses developed systems of resolving personal and policy conflicts within the Capitol building, often leaving even the media's political reporters mystified. At one point, when even the tiniest gifts from lobbyists were forbidden, legislators might eat together and entertain themselves cheaply across party lines without any sense of awkwardness. They didn't necessarily discuss legislation over salads, but they acquired a sense of how their colleagues valued daily decisions in ordinary life. That familiarity created trust in the committee rooms, where the major issues of legislation are really addressed. Trust is an essential element in persuasive discourse.

During her tenure in the legislature, my wife guided the formation of the Iowa Peace Institute. It was at the outset directed toward international affairs and was partially designed to become an entity for the training of diplomats in resolving conflicts. It later has redirected its energies to more local conflicts. Although it still maintains interests in world peace, it is more often concerned with conflicts among local governments in Iowa. Much of its effort is directed to training local people to manage conflict, especially in schools. It runs widely praised workshops for the Department of Education to help train both teachers and students how to work out the usual school conflicts. One of their favorite anecdotes is of a sixth grader who used the techniques she learned in school to help resolve conflicts between her parents.

Those same years were extremely difficult for farmers, and loans on farms were being foreclosed all over the state. That had led to the creation of a farmer/creditor mediation service to provide a context in which creditors and farmers alike could work out plans that might keep farmers working and investors getting paid without resorting to bankruptcy. Certainly it was not good for farmers to be put out of business, and banks didn't really want to have to find some ad hoc way to keep the land productive. These two quasi-governmental bodies were designed to turn the energy of conflict into positive programs. My wife, having supported both of these groups as a legislator, became particularly interested in their procedures and helped define their functions so they might work coordinately. She retired from the legislature and entered a graduate program to study conflict resolution, so I became an onlooker of the field in a different venue.

As a culmination of her theoretical studies, she examined in detail the processes by which a major governmental reorganization bill had passed our legislature a decade earlier and was signed by the governor despite the initial anxieties of state agencies being relocated or even abolished. Brought to bear on this one bill were all of the usual conflicts between parties and houses, between the executive and the legislature, and among all sorts of special-interest groups around the state, so it is an excellent example of how public conflicts might be altered into a generally accepted reform. In her analysis of verbal strategies, she found it helpful to borrow several of my linguistic and rhetorical reference books, and I was struck by how the concerns of conflict resolution at the practical level were examined in different terminology by rhetorical theorists.

Because mediators have as many variations in their theories as do rhetoricians, and for my purpose here I need only a representative presentation, I've chosen one book, the second edition of *Getting to Yes* by Roger Fisher, William Ury, and Bruce Patton of the Harvard Negotiation Project, as a benchmark. I use it because it is readable, it is readily available in a Penguin paperback, it is much favored in short courses and on business reading lists, and it is clearly mainstream.

The authors support "principled negotiation." They describe their method under four headings: (a) Separate the People from the Problem; (b) Focus on Interests, Not Positions; (c) Invent Options for Mutual Gain; and (d) Insist on Using Objective Criteria. These are not absolute groupings; they are convenient for writers in presenting the process. In effect, the Harvard people organize and describe practical situations of conflict in such a way that standard freshman textbook advice on rhetoric can be enriched through reinterpretation.

Separating people from the problem is in simple terms examining how the speaker and the audience affect the analysis of the external subject. (In any discussion the roles of speaker and audience alternate, and one might use the terms *writer* and *reader* as well. Here I'll stipulate that these are all synonyms for the sides in a conflict.) People have interests and limitations that tend to skew their views of what exists between them. Some of the interests and limitations are relevant to the situation at hand in a judicial sense; many are not but still affect the possibilities for resolution. The problem for any mediator is to discover the difference.

When a landlord files for back rent in a small-claims court, the issue in a legal sense may be simple. The lease says payment is required. That the tenant is broke, the house needs paint, the furnace does not work well, the landlord is black, the tenant has a child requiring medical treatment in the hundred-thousand-dollar range, the previous tenant has been bad-mouthing the landlord, the landlord has fifty highly profitable rental units, the tenant is white, the landlord has never expressed sympathy about the child—all may be judicially irrelevant and yet at the heart of the conflict. A lease may cover some of the side issues, or it may not, but it probably leaves out what causes the conflict. Quite possibly a mediator will want to probe what source and level of anger really brought these parties to court, and quite possibly it will turn out that neither side recognized the humanity of the other person. We all have troubles. Small claims officially are about money, for one must bring up money in order to file a claim, but the actual issues may be quite different. In order to deal fully with the claim at hand, the mediator must go far beyond what a judge would be required to observe. One might wonder what a mediator would have done for the merchant of Venice, for the issue in the play is surely not a matter of ducats.

As a practical matter, the mediator may have to meet separately with the antagonists in order to let them vent some of the anger, but sooner or later the participants need to discover more about their opponent and themselves in order to isolate what is the real issue, the problem. One needs to see from the other person's point of view even if one clings to one's own vision, although usually double vision alters both attitude and claims. "Yes you did, no I didn't" is ultimately an unproductive discussion. In some ways even talking about "the problem" focuses the discussion too soon. Still, as school debaters learn, establishing a need for a change, a problem to be solved, is the first step in debate. All too often proponents have a solution before the problem is defined. Put in a new stop light before the traffic patterns are described, lock the doors of the school before safety hazards are evaluated, cut taxes before the value of services is assessed, give more tests to improve the quality of the schools. The series can be extended indefinitely. Usually, proponents have mixed motives for their making a proposal, so those need to be sorted out, and most problems are many faceted and require analysis.

That leads to the second part of the method: Focus on interests rather than positions. People who have decided on stop lights or metal detectors or tax relief or lower taxes often fall in love with their positions. Probably they really want orderly traffic or a safe school or good service at a fair price or sound education, but having moved too soon to a solution, they have a hard time defending or even identifying their own interests because they are too busy supporting their proposal. They haven't really defined the problem. Stepping back to discover what they really want will enable them to examine solutions more calmly, so the speaker who would help them must adopt a persuasive strategy that supports a wider context. Even the landlord might find that in the interests of helping those in trouble, he can delay rent payments, and the tenant might be more ready to find some way to meet his legal obligations if he felt the landlord cared about how the child was doing and would help make the situation bearable.

Interests and motivation are clearly related concepts. Although we talk of motivating listeners, we really mean that we will discover the motives they already have and explain our position in terms of those motives. The mediator seeks the motives—interests—of all sides of a dispute in hopes of finding common ground. Persons wanting Medicare to pay for prescriptions probably are interested in staying alive without going broke. Most people in the culture would accept the legitimacy of those motives. An opponent of government aid might have a different position—get a better job, for example—but in the broader context, opponents might be willing to recognize greater complexity in the situation and the need for alternative positions.

The third step in this kind of problem-solving is to seek additional options. One can hear in this advice all of the procedures for the brainstorming techniques popular thirty or forty years ago. A group of interested people was directed to offer solutions to some problem as quickly as possible, and the participants were instructed not to object to other ideas. No matter how crazy the idea, it was allowed a place on the list of options. Even odd notions might stimulate others to think "out of the box," to take up a later buzz phrase. To be sure, systematic people might develop scales or paradigms to exhaust the range of possibilities. Practical rhetoricians and problem-solvers may have many inventive techniques, but the point is to free oneself from the tyranny of position. The Harvard people, too, offer many techniques, and the standard school rhetorics from Aristotle onward are packed with them. They constantly lead one to re-see the problem and its solutions. If we think of rhetoric as merely enforcing a position already adopted, we limit its usefulness in mediating among various human interests.

When a wide range of options are laid bare, choices have to be made in a manner that satisfies the parties. The final step of the mediation procedure is to agree on objective criteria for identifying the optimal solution. Any house builder will recognize the importance of bidding specifications. Building codes take care of many details, but a building contract ordinarily has many other standards usually quantified to forestall later conflict. We are encouraged to objectify the evaluating process while reconciliation is still cheap. Most conflicts are less tidy. Writing objective specifications for an academic job notice challenges the adept rhetoricians. The solicitation must be legal and relevant, it must attract the qualified people, and it must define the job that really has to be filled, but it probably cannot be quantified. "Objective" has to mean more than quantification or simple categorizing. Terms like *comp/rhet specialist* turn out to be squishy. One can stipulate a meaning by listing how much training of what kind is required, but doing so cuts off options for recognizing some excellent but unusual person. We can argue that a new garbage dump needs to be accessible and inoffensive—objective-sounding standards—but almost by definition if a spot is already accessible, it probably also has people who will be offended by garbage.

Objectivity in disputes is a goal, not a certainty. We can struggle to eliminate all of our reactions to the person and recognize our own biases; we can try to recognize our interests in demonstrable terms, but in the end human value systems are not identical. Otherwise, there would not be systems—plural. We do what we can to make issues overt so that the parties to the dispute can see what complications make solutions difficult and find what they really have in common, but we probably will find that their interests really are not identical. Still, their differing interests may be satisfied by one of the new options they discovered in brainstorming. Perhaps the dump can be designed as a tidy recycling plant. We work to eliminate unnecessary disputes over prejudged positions, and then we consider what we do to satisfy the remaining interests.

Sometimes we are left with interests that cannot be resolved by negotiation. That is why we have laws—ultimate objective standards whether sensible or not. The legal system comes into play when principled negotiation fails because society has to have a way to get on with its business. That is why we punish protesters who break the law even if their cause is good. In Ming China going to court was in itself shameful because it represented a failure of community. But law west of the Pecos was the gun. Law in international affairs has often been war. We prefer to recognize law that has been jointly and rationally arrived at, but in the end, even that kind of law is force. We accept the force of law because a lawless state is unacceptable to us, and when we are thwarted, we probably undertake to change the law or redefine our interests. Indeed, people who are coerced by law at first often eventually come to believe in the law that they opposed, as demonstrated in the history of civil rights legislation. As Hamlet tells Gertrude, "Assume a virtue if you have it not," because practicing virtue will eventually make you virtuous.

Having said that, I acknowledge that in our world obeying the law is largely voluntary. We could not possibly have enough police to enforce all the laws except that most people follow the rules. Force is used to keep extreme behavior controlled, and some scofflaws don't want to pay the penalty and do a good bit of whining when submitting to force. I also concede the existence of suicide bombers. At some point rhetoric and conflict resolution just don't work. Compromising in order to get a settlement or even obeying the law probably is a short-term orderly solution buying time, because a person who is unsatisfied in bowing to the inevitable probably will try again. Still, delay allows time for more options to be suggested, and interests to be revalued. A writer knows that a text is never finished; it is merely abandoned to a publisher's deadline. A negotiation is probably never finished even if the present conflict is resolved because situations change. That is the nature of processes. One needs strategies for delay. In addition to providing a general strategy in the four steps of principled negotiation, Fisher et al. devote many pages to dealing with intransigence and provide all sorts of examples for analysis.

For those who try to define rhetoric, this excursion into conflict resolution is worth extending. The word *rhetoric* does not have much standing in the world beyond itself. One hardly hears the word without "mere" in front of it. Rhetoric remains central to society by whatever name it appears because it describes and prescribes fundamental behaviors. From the beginning its roots have been practical, so practitioners operating on tacit theories have demonstrated what actions best serve individuals in a society. If rhetoric is persuasion, it is so because human relationships in themselves—interactions—imply persuasion, actions of people on one another. As the Existentialists observed, choice and not-choice are equally implicit recommendations for behavior. How any generation decides to focus its awareness of these interactions is revealed in the terms it chooses, and just how conflict resolution, mediation, and negotiation are revealing terms for much of what is covered in rhetoric, and the discussions often sound like mere descriptions of what people do.

The reorganization of state government in Iowa illustrates how the practical techniques exist apart from discussions of theory. Although there was general agreement that the state had too many agencies, commissions, and boards in too many forms reporting (or not reporting) in illogical ways, the problem aroused little more activity than oratorical position-taking. The governor had a vested interest in more efficient administration since it would make him a more powerful governor. Two or three possible gubernatorial candidates in the legislature had interests in creating an executive more rewarding to them. The general public had an interest in more effective governance as well as an interest in broader representation. Individual agencies had interests pushing their public agendas and their private ambitions. These interests were not in harmony.

The appropriate senate committee used the situation primarily for seeking media attention and for letting various special interests air their self-serving positions. The house committee, on the other hand, conducted systematic public hearings and invited agencies to present fairly orderly cases for what they did, for making explicit what their interests were. In a sense the house committee was co-opting some of the parties by making them share in the process. They also heard expert testimony from various specialists in government primarily to discover options for governmental organization and criteria for measuring effective systems. That is, they tried to separate personal stakes from the issues of governance and find objective ways for examining options. On the whole, news gatherers found these procedures lacked drama and were too academic and too polite to require much coverage. They chose most often to interview senators.

Early in the hearings, the committee began regular discussions with the governor, who was not of the party that held the majority in the legislature. In consultation they began to work out specific plans to fit principles offered by the experts. As the sections of the overall plan were worked out, those affected at the working level were invited to see how the new plan would better serve their individual interests by clustering for mutual support departments with similar interests. The governor saw simpler lines of control and responsibility. The legislative rivals to the governor came to see that they would be better served claiming credit for the new ideas and later for governing (maybe) with them in place. People on the senate committee observed that most of their complaints were addressed by the new plan and that they could share credit if they signed on. Wider representation was assured by rules requiring gender-balanced governing boards and legislative oversight of appointments for policy positions. The plan was devised so as to meet the needs (interests) of all relevant groups. To be sure, one or two anomalies were allowed because some agencies resisted being grouped with other agencies; one thought it had power that it didn't want to share, and another just didn't like the company. These variations were minor, and by conceding that great improvement is possible without abstract perfection, the proponents saw that the bill was passed with minimal posturing on the floor of the legislature.

Here was a reorganization of the executive branch that reduced 138 agencies and departments to 22 and rearranged them according to missions and levels of contact with the governor. The idea was barely considered and not given much chance of passing at the outset of the session, and yet by accepting reasoned and principled negotiation, the legislature adopted the specific plan. More than a decade later, one sees that it has worked well in drawing good people to serve in government and in getting work done. Perhaps some legislators did not quite realize the implications of what they were voting for. For example, some probably did not realize that requiring gender-balancing in policy committees would be important, but it has probably altered the face of government and draws very little complaint. Indeed, very few efforts to alter the original plan suggest it has worked well.

Although I believe Iowa can be proud of its reorganization, that is not the point here. The people who accomplished this potentially upsetting change were not theoretical mediators or rhetoricians. As politicians they probably had some practical standing as both, but we all have observed that politicians often exert the rhetoric of force. These people took a broader view of discourse, one more in accord with deliberation as Aristotle seemed to understand it. They weren't really working toward compromise—giving up something you don't really care about in order to get something else that is important to you—but sought through discussion to devise a plan that could meet the interests of all involved. They tried to be open to solutions.

In 1975 Jim Corder in a casual-seeming essay for *CCC* wrote about what he learned in school. He draws upon the immediate experience of writing along with the students the papers he assigned in class as well as on his professional learning. He offers in conclusion some laws of composition rather randomly numbered. In his twenty-seventh law, he puts the practical problems of conflict in rhetorical terms. "Invention invites you to be open to a creation filled with copious wonders, trivialities, sorrows and amazements. Structure requires that you close. You are asked to be open and always closing." What a conflict, what tension! Anyone who writes knows how beguiling are the side issues, but the conflicting need for closure contributes to the energy that produces text. If one is lucky, there is language to bring humans together for a time. For professionals in the rhet/comp area who sometimes denigrate literature, it is worth recalling that Corder preferred an academic universe that included both on equal standing. Perhaps good administrators have to recognize contexts greater than the one they direct in order to recognize how interests are related. In a slightly different academic arena, rhetorical practice precedes rhetorical theory and should be part of education of all people, but attention to theory can enable a person to choose the better practice. Both, not which, is guide to curriculum makers.

Rhetoric is at least a metaphor for all human relationships and therefore is a system for combining and resolving conflicting interests, but it can also describe the techniques of separation. In a world of force, rhetoric offers the tools of repression and resistance. In a civil world, rhetoric gives guidance to those who look to a pooling of interests. As the human capacity for exerting force becomes increasingly horrifying, survival depends more and more on our ability to negotiate agreements without sacrificing our sense of self. We need to be expert in using the skills of reconciliation and in recognizing what lies beneath the rhetoric of hostility. There is no such thing as an amorally neutral composition course.

WORKS CITED

Corder, Jim W. "What I Learned at School." *College Composition and Communication* 26 (Dec. 1975): 330-34.
Fisher, Roger, William Ury, and Bruce Patton. *Getting to Yes: Negotiating Agreement Without Giving In.* 2nd ed. Penguin: New York, 1991.

11
Rhetoricians at War and Peace

Elizabeth Ervin
University of North Carolina, Wilmington

Author's Note, December 2001:

It has been over a year since I have written this chapter, and the United States is once again at war. It is a troublingly indeterminate war: "Operation Enduring Freedom," it's called, and that "-ing" bothers me, suggesting as it does a long, maybe even perpetual, period of violence, suspicion, and circumscribed civil liberties. Having been given the chance to update my chapter in light of these events, I am tempted simply to retell the story of September 11, 2001, as one of close calls and witnessing, of the guilt of nonparticipation promoting entrance into a vast universe of obligation. After experiencing the outpouring of generosity and rage in the wake of the terrorist attacks on the United States, who can doubt the potency of these themes?

But this story is still unfolding, and I would be foolish to speculate about its direction. Moreover, to treat it as merely a series of current events is to miss an opportunity to think about it as rhetoric—that is, to test the ideas I have discussed here now that the stakes have been raised. Probably most of us have followed with some ambivalence debates on the appropriateness of expressing dissent in a time of crisis (is it more patriotic to tolerate these voices or to suppress them in the name of unity and "national security"?) as well as on the perceived moral relativism of contemplating why people around the world resent, even hate, Americans. Perhaps Corder's vision of a "great, unbounded possible universe of invention" seems naive in light of our changed circumstances. But as Lisa Ruddick suggests in an essay in The Chronicle of Higher Education, *these new circumstances may make Corder's ideas even more compelling as they mark an occasion for us to "re-create a sense of meaning" in our personal and professional lives, to meet one another "on some shared human ground" (B8, B7). Ruddick fortuitously answers the question that Mary Rose O'Reilley poses at the beginning of this chapter by asserting that all academic inquiry—in particular, English studies—can represent a practice of understanding "what sustains people" (B7). Toward that end Ruddick recommends that we embark upon "a complex exploration of the art of listening that is one of the creative forces in the world, a force moreover that our species*

185

would do well to cultivate if we want to have a good chance of surviving" (B9).

It is impossible to predict how our individual or collective rhetorical sensibilities will be affected by the events and aftermath of September 11. But here is what is possible: affirming through practice a rhetoric that challenges us to contemplate our most deeply human concerns and listening—not merely surrendering a turn, not formulating our next comment as the other person speaks—as we grope toward wisdom, agency, and love.

❧In *The Peaceable Classroom*, Mary Rose O'Reilley asks a question that lays bare every anxiety of our profession: "Why teach English?" It's a question that nearly all of us have pondered at one time or another, perhaps even avoided, because its implications are so profound. But as O'Reilley puts it, "If we don't have a good answer . . . there is lots to do elsewhere in the world or down the back streets of our cities" (62). O'Reilley's own answer takes me by surprise. "[F]inding voice is a socially-responsible political act," she asserts, and assisting others in this effort is akin to mounting a mass protest against "the suppression of the personal and idiosyncratic" that generates—and signals—an atmosphere ripe for war (62, 59). Her point is as radical as it is mundane: Those who have access to their own voices, to a language that authorizes them to act on their convictions, will resist violence and seek to create something better in its place; those who do not will remain vulnerable to manipulation and oppression (59-60).

This impulse to "create something better" is easier theorized than done, however, in part because our cognizance of war is itself increasingly theoretical and detached (and really, who would want it otherwise?). As Michael Blitz and C. Mark Hurlbert observe in *Letters for the Living*,

> [O]ur students are growing up in a different age—a different war-consciousness. Their sense of war and peace is shaped as much by microviolences as by the Big War scenarios. What do they know of peace? They know it as a term in their history books; they know it as a chip of rhetoric, a sound bite in political speeches, a retro symbol of the '60s. . . . And so if they come to us for peace, they come to us for the resolution of an image that is blurry, abstract, perhaps beyond them (perhaps beyond us). (66-67)

It's not entirely misguided to focus on what students do or don't bring to their learning, but we must admit that our own war-consciousness is very much at issue here as well. Unlike Blitz and Hurlbert, I never "watch[ed] friends and relatives fly off to Vietnam to be shot by enemies and friendly fire" (66-67); my experience of war is limited to viewing Operation Desert Storm and other organized violence on TV. War, in short, remains blessedly beyond me. Does that make peace an equally inscrutable abstraction?

We might usefully contemplate this dilemma within the context of our discipline's own recent history, one in which the complex connections among rhetoric, war, and peace were understood as matters literally of life and death. Kenneth Burke's views on the subject need hardly be recounted here, but it's worth remembering that I. A. Richards, a contemporary of Burke's, developed his ideas in part out of concerns about wartime propaganda and nuclear proliferation. Ann Berthoff reminds us, for example, that in Richards' lexicon, *choice*—purposeful action—advanced the "arch-purpose" of planetary survival, and "the purposeful choice of words serve[d] as a pattern for all choices" (278). Choice is of course always circumscribed by context, and when it comes to war, one of the more immediately influential contexts is also, among academics, one of the more inimical: military service. Such has not always been the case, however, and in this chapter I investigate the possibility that military service has inspired several of our own contemporaries to enact rhetorics in which the threat of violence is less imminent and the possibility of peace more imaginable. I begin by exploring the phenomenon of close calls—where individuals are unexpectedly spared the experience of combat—and proposing a reading of those rhetorics that appear to be inspired by such experiences. Next, I turn my attention to Jim Corder, using details from his life and work to amplify this discussion, showing how invention and revision can serve as instruments of peacemaking in a world of shifting personal and physical landscapes.

RHETORIC, TESTIMONY, AND THE GUILT OF NONPARTICIPATION

In this first section, I attempt to isolate the complex dynamics of relief, guilt, and obligation typical of responses to wartime close calls and examine ways that they manifest themselves in ideas about language. My primary focus will be Wayne Booth, for two reasons: His military service coincided with the experiences of Holocaust survivors, whose narratives, I suggest, offer a template for understanding close calls; and his extensive body of scholarship is consistently concerned with using rhetoric for the purposes of fellowship and mutual agreement. I draw the biographical details from Booth's most recent book, *For the Love of It*, a meditation on pleasure and wisdom organized around his lifelong love of music and his efforts to learn to play the cello as an adult.

In the fall of 1944, Booth tells us, he was a "belated draftee, a 'clerk-typist' . . . being shipped toward the European theater" after serving as a missionary for the Mormon Church (32). On the last day of boot camp, his sergeant informs the company that instead of being clerical workers, they are all going to be riflemen, and Booth begins "to believe that I would die in this war, a war in which I saw our role as necessary and *perhaps* noble" (32). Feeling ill-fated and tragic, he consoles himself with classical music and the prospect of his own heroism. Some weeks later, on a cold, wet morning, members of Booth's company line up; one by one their names are called, and one by one they dutifully climb into trucks to be sent to the front lines. Booth hears his own name, but when he steps forward, he's told that he's going back to Paris to be a typist after all. Writing in his journal, he describes his ambivalence toward this turn of events:

> My eighty-word typing speed had saved me! Yet somehow I felt more lost . . . than when I had been heading for doom. Riding back to Paris in a rough, cold army truck, I thought long and hard about the unfairness of my rescue. Many of my buddies had wives and children back home, while I . . . well, all I had was a fiancée who could easily find another man. Gradually as we rolled along I came to an absolute decision about God, for the first time since my childhood orthodoxy. It was not now just a strong suspicion but a firm rejection: any God who could play an unfair trick like this on those miserable buddies was no God at all. He had died, and along with Him somehow the music faded. I had been singing it as my sacred death dirge; now it had no special place. The sacred had fled, while I was being "saved." (34-35)

Booth's religious convictions were irrevocably altered by this experience, as were his attitudes toward music. While this latter change might seem inconsequential, it was to have a profound effect on his later thinking. For in the aftermath of his reprieve from combat, learning to play the cello no longer held the naive emotional significances, but rather came to evince "sheer physical struggle . . . performed with love but without hope of total success" (35). In other words, his experience led him to accept that because neither noble intentions nor cosmic justice inevitably win the day, actions must be performed in spite of their difficulty, and for the sake of love and communion rather than mastery.

Booth's sentiments are echoed by other rhetoricians who have experienced wartime close calls. Lad Tobin, for example, describes a similar response to his successful efforts to evade service during the Vietnam War. After receiving a letter from the Selective Service during the summer before starting college, Tobin inquires into the possibility of gaining conscientious-objector status, attempts to get a psychiatric exemption, and solicits letters from various family doctors attesting to his "severe asthma" ("not mentioning of course that it was not so severe as to keep me off the wrestling and cross-country teams," he adds wryly) (81). This last scheme seemed so foolproof that Tobin was completely unprepared for the moment "when I received a letter in my college mailbox notifying me to report in three weeks for an induction physical" (76). His comfortable middle-class upbringing had, he admits, led him to believe that he was invincible to hardships that routinely affected others, and in fact it was this upbringing—buttressed by medical specialists and protective parents—that did ultimately keep him from serving in the war. He writes:

> Finally, at the very end of the day, I waited in line to meet with an army doctor. "You have passed all the exams," he told me, "and will be inducted into the U.S. Army next month. Is there any information that you have that would disqualify you [from] military duty?" With desperate gratitude for the question, I handed him the letters and began to tell him about how useless I'd be in the jungle, about how long it takes me to catch [my] breath after a jog, about. . . . (81)

Eventually, Tobin persuades the doctor of his unsuitability for combat and is excused from military service. But despite the self-described desperation with which he sought to avoid the draft, Tobin doesn't celebrate his good fortune. Instead, he says, "all I felt was numb" (82).

The combination of relief, despair, shame, and loss of faith that Booth and Tobin express in these passages bears a conspicuous resemblance to the survivor's guilt experienced by World War II concentration camp survivors: a lingering sense that they had been spared at the expense of others who had died, that they had not done enough to resist or help others, that they were less deserving of life. A variation of survivor's guilt has been characterized by Ellen S. Fine as "the guilt of nonparticipation," experienced by Jews who emigrated at the beginning of the war or who hid or "passed" during the war, Europeans and others who did not fight, and even children of survivors who feel remorse for not being born in time to save their parents from suffering (192).

Holocaust scholarship contains many representations of this guilt. A particularly poignant example comes from the oral history of Margie Nitzan, whose family emigrated to Italy in 1935 when she was still a child and lived there during much of the war. During this time Nitzan and her sister found refuge in a Catholic orphanage, where they were indoctrinated to believe that Jews were inferior and willingly converted to Catholicism (Gurewitsch 124). In 1946 Nitzan moved to New York City with her mother and siblings; ten years later she reverted to Judaism and began reading the *Aufbau*, a German-language newspaper published by and for the Jewish refugee community. Although she had previously believed herself to be unaffected by the war, Nitzan says:

> I ruined my head reading everything about the concentration camps. I used to think about it and hear screaming, and I just relive the whole thing. I went through it, and I was out of my mind completely. I had that syndrome: why was I living and the others died? What did they do wrong? I used to feel that I was a German and I killed the Jews. And that's really . . . I didn't do anything. (Qtd. in Gurewitsch 126)

So profound was Nitzan's guilt of nonparticipation and the accompanying self-loathing that she actually began to consider herself a Nazi, an identification that haunted her the rest of her life. At the time her oral history was published, Nitzan was living in a psychiatric facility and participating in a community literacy program; one of her poems, "Margie Nitzan Autobiography," is included at the end of her narrative. In it Nitzan details her painful conversion and concludes with these lines: "Until this day I regret it, It / almost cost me my / life" (127). Another close call.

Nitzan is typical in her efforts to revise unhappy war experiences through positive actions (reclaiming ethnic identity, writing poetry), and as such, her account suggests ways in which the guilt of nonparticipation can provide an impetus for composing alternate narratives about war and peace. We can discern similar patterns in Wayne Booth's rhetorical theories, as when war and religious skepticism are invoked—often in tandem—as metaphors, as historical realities, and as inducements to challenging the inevitability of violence and conflict. For example, Booth writes in *Modern Dogma and the Rhetoric of Assent*:

> We are a society groping for meaningful affirmation, for intellectually respectable assent. The old faiths seem shattered; men have been told again and again that they must either leap into faith blindly or abandon faith entirely; the mind doubts, the heart may—if it doesn't care about being a bit stupid—affirm. But we know what we know: affirmation, assent to any but the devil's camp is not going to be easy for any man with his eyes open—in the age of Belsen, Dresden, Hiroshima, Vietnam, and the thousands of daily horrors that some [of us] seem almost to enjoy recounting. (200-01)

Thus it makes sense to regard Booth's wartime experience and his resultant crisis of faith as significant intellectual influences, particularly on his "rhetoric of assent": the rational use of language for the purpose of saying "yes" to the world and, in so doing engaging in the "communal building of selves" (*Modern* 196). Booth's love of music and his rhetorical studies manifest this assenting stance:

> [B]oth fit under a special notion of "love," the belief that in our efforts to *communicate better*, whether in words or music, we often are, and always should be, striving to meet "the others," the "other" that is now fashionable. . . . Human love, human joining, "critical understanding" as a loving effort to understand—that has always been at the center [of my endeavors]. (*For the Love* 121)

This description of the aims of rhetoric is consistent with characterizations of Holocaust writing as "act[s] of love and the only steady avenue of communication" between those who endured these events and those who did not (Horowitz 292). And if World War II represents a rhetorical exigency for both survivors and nonparticipants, then Holocaust testimonies and Booth's rhetoric of assent can be read as analogous responses.

Although Tobin managed to avoid military service, his ideas likewise find illumination within the context of Holocaust narratives. His personal disclosures are part of a larger exploration of why students repeatedly write about "loss, anger, confusion, and grief" (73), and moreover, why they insist on writing happy endings to such essays. In the end his question changes from "Why do they write about these things?" to "How could they *not* write about them?" (83). Narrating and revising difficult experiences like close calls may, in short, be intellectual and spiritual imperatives. Thomas E. Recchio, another war veteran, calls this impulse "the critical necessity of 'essaying.'" As a 21-year-old Marine, Recchio was stationed in Japan during the Vietnam War, where he discovers Russian literature during an extended period of boredom and inactivity. He reads voraciously until he is released from the military, at which point he turns to writing. Referring to himself in the third person, Recchio says:

> He begins to understand his life as a kind of essay, a continual restless effort to compose a self in relation to his experience of the world and of language. . . . He feels fortunate, even blessed, to have found a kind of work that seems to offer possibilities for wholeness, where his personal and professional lives can become one. He recalls the alienation of his military experience when wholeness was an illusion conceivable only through numbing both body and mind. (221)

Although perhaps unaware whether he would enter into combat, Recchio does experience the numbness and disaffection that others have described in response to close calls, as well as the familiar impulse to revise. Adopting Graham Good's definition of the essay as "an act of personal witness" (219), he claims that "[t]he essay bestows on us the authority to write and rewrite in an effort to understand and, through understanding, to remake ourselves, our work and our lives" (233-34).

Recchio's comparison of "essaying" and witnessing is fitting but not unproblematic within the context of Holocaust scholarship, which draws a critical distinction between *witnesses* and *bystanders.* Departing somewhat from the legal definition, Victoria J. Barnett defines bystanders as those who are silent and passive in the face of others' suffering, perhaps convinced that they are powerless to intervene in events or institutions that seem much larger than themselves. Witnesses, by contrast, inhabit what Helen Fein calls "the universe of obligation" and, moreover, expand the boundaries of this universe to include those outside their own group "whose injuries call for expiation by the community" (33); although they need not be heroic, they take action on behalf of others, sometimes at great personal risk. While indifference to suffering "utterly dehumanize[s]" both victims and bystanders, according to Barnett, the actions of witnesses do the opposite (9). Because witnesses don't merely feel sorrowful in the face of horrors such as genocide but act in response, they earn the right to construct testimonies of their experience, a status they share with survivors.[1]

These definitions beg an important question: Do nonparticipant testimonies represent rewards for actions or actions themselves? Holocaust scholarship is equivocal on this issue (is there a statute of limitations on responsive or redemptory action?), but rhetorical scholarship is not: Rhetoric is a form of action. Informed by the concept of praxis, which upholds a dialectical relationship between action and reflection, many rhetoricians believe, with Paulo Freire, that "to speak a true word is to transform the world" (75). Thus while Booth, Tobin, Recchio, and (as we shall see) Corder may have been ambivalent nonparticipants in war (whether by virtue of fate, stratagem, timing, or choice), they voluntarily occupied a universe of obligation that necessitated their participation in peacemaking. Though removed from combat, in other words, they were not merely bystanders to others' suffering. Rather, in response to the injustices of war—including their own close calls—they constructed and enacted affirming, humanizing, self-actualizing forms of communication. Their rhetorical theories might therefore be read as testimonies of witness.[2]

MUSTERING THE DISCOMPOSED SELF

I turn now to Jim Corder, whose writing epitomizes witness rhetorics borne of presumed wartime nonparticipation. I'll begin with his book *Yonder*, a collection of interrelated personal essays that weave together themes of change, memory, nostalgia, and uncertainty against the backdrop of personal geographies and the constructed self. Central to these essays are three significant events in Corder's life: moving with his family from the small town of Jayton, Texas, to the city of Fort Worth in 1939; the disillusionment he experienced in the wake of World War II and the Korean War; and finally, his divorce and subsequent estrangement from his eldest daughter. I should emphasize that although my ultimate interest lies with Corder's ideas about rhetoric and although *Yonder* was published some years after his most influential scholarly work, I think it is appropriate to open with the autobiographical, for it gives us insight into the motifs that Corder himself considered important. Throughout these essays, for example, we recognize the phrase "perhaps we will study war no more," versions of which appear so often in Corder's writing as to represent an intellectual mantra.

Corder's life as a soldier was spent in Germany from 1951 to 1952 where he was an Army sergeant and, according to his widow, Roberta, "had something to do with tanks" (March 29, 2000). Although the details of his military service are largely unknown to me, it is clear that he did not go into combat: "I have not been to war, though near," he writes. "I have not watched when mother, father, sister, died as bombs fell or gas hissed" (38). Still, he asks to "be released a little from culpability from my own crimes of commission and omission," suggesting that he feels some sense of guilt for the consequences of his involvement, or lack of involvement, in various conflicts (38). This guilt causes him to rethink many of his old assumptions about war, as he reveals in a later essay:

> I regret that I was ever attracted by war stories, and yet I have turned back, time and again, and not so long ago, to accounts of the Battle of Verdun . . . wondering whether or not I could have measured up to that hellish testing. . . .
>
> [Growing up] I was sorry that I wasn't old enough to go to war. It didn't seem right that others had to go while I stayed, safe and cool in the shade on Cleckler Street. I regret that I was ever attracted by war stories, and I regret that I ever thought war to be the appropriate test for manhood.
>
> But it's not easy to give up your war stories; it entails regretting your life. . . . [I]t's hard to change some habits of behavior and thought. But I'm going to try. I'm not going to study war anymore. Or competition. Or manhood defined in those old terms. I may not make it. ("World" 26)

Thus we see the chain of inaction, regret, and resolution to "do otherwise" typical of close calls. But even as Corder rejects war stories as propaganda that exploits young men's dreams of heroism—and although he claims never to have measured up to them, anyway—he acknowledges their appeal as means of validating a life that otherwise might appear to be characterized by an undistinguished record of nonparticipation.

The fact that Corder served in the Korean War is itself suggestive. While perhaps never in immediate danger of deployment, as a member of the American Occupation Force, Corder certainly lived among the ghosts of that earlier war, both living and dead, including that of his own brother (see "Argument" 27). Roberta confirms, moreover, that he "was constantly reading and writing about" and "deeply affected by" World War II (March 30, 2000). This influence is visible throughout his work (for example, in his frequent references to narratives, oral histories, films, and so on related to World War II and the American Civil War, in which he sees articulated his own postwar sentiments), and indeed, the Holocaust emerges as a central trope in his writing. While Corder emphatically does not equate his own personal sorrows with those of concentration-camp survivors, he does reflect on the "candle's worth of holocaust" that he has experienced, in particular, the violence of being uprooted from his home and feeling "a little lost ever since" (*Yonder* 41, 39).

This sense of displacement runs throughout Corder's work, as illustrated by his frequent reference to the diaspora: site of both exile and provisional refuge. While Corder acknowledges the term's ethnic associations, he suggests that the diaspora is also experienced "in intense, sometimes agonized, personal ways" through feelings of dislocation and yearning for what is lost (22). He writes:

> During these later years I've gone back [to Jayton] to learn the
> name and look and situation of things, but the things I've
> looked for aren't there, and when I look at my maps, I'm not
> there either. Once I was there, and once my people were
> scattered over parts of three counties. . . . Now they are gone.
> The wind passes over the places where they were, and they are
> gone, and the places where they were will know them no more.
> (39-40)

Corder's confidence in his own identity seems to disappear with his ancestral home, leaving him feeling rootless and unstable and compulsively checking to see if he still exists. In kind if not in degree, his anxieties resemble those of Holocaust survivors, many of whom, as Lawrence L. Langer explains, attempt to heal their ruptured lives by finding a logic that integrates their past and present. It's a futile endeavor that can actually diminish rather than restore the self, according to Langer. He believes that instead of imposing artificial order on the chaos of witness testimony, we must accept "unreconciled understanding" as evidence "that we can inhabit more than one moral space at the same time . . . and feel oriented and disoriented simultaneously" (201). *Yonder* is filled with phrases that reveal Corder's acceptance of this same paradoxical truth: "[W]e can mostly speak only in the language of the day, which is still partly yesterday's language, though it can't be, and must be tomorrow's language, though it won't be" (24).

While the diaspora can arouse loneliness and unrealizable expectations of wholeness, Corder also regards it as a site of reflection, empathy, and second chances. The fissures inherent to the diaspora provide opportunities for both invention and revision and, moreover, can reactivate the displaced person's efficacy as agent of these processes: "I think identity wavers out there elsewhere, and time breaks, and landscapes shift and erode. Perhaps we all wander, lost in migration, trying to find ways to transform ourselves, trying to fix ourselves, to repair the world" (*Yonder* 43). Corder's language here echoes the Jewish expression *tikkun ólam*, "repair of the world," which has assumed a special urgency in the years since World War II. The combination of Corder's personal traumas and postwar guilt apparently led him to embrace the impossible, necessary effort of reparations—revision, in rhetorical terms—which even discomposed selves can render and receive, and in which resides the possibility of restoring trust in human compassion (see also *Yonder* 89).

In Corder's view, however, revision does not inevitably pay off, as some composition textbooks would have us believe. "Why would we suppose that revision is always good, or even possible?" he asks. "Perhaps I can guess. We always had the hope of getting it right, sometime. We always had the knowledge that we mostly don't" (*Yonder* 91). In other words, some texts—lives, testimonies, landscapes—simply resist "fixing" and remain confused and incongruous. Thus redemption may not be achieved through revision but through invention, which in

Corder's cosmology is a radically different, and more optimistic, concept: a "great miracle" that can make it possible "to see each other, to know each other, to be present to each other, to embrace each other" ("Argument" 29, 25).

> By its nature, invention asks us to open ourselves to the richness of creation, to plumb its depths, search its expanses, and track its chronologies. . . . Each utterance may deplete the inventive possibilities if a speaker falls into arrogance, ignorance, or dogma. But each utterance, if the speaker having spoken opens again, may also nurture and replenish the speaker's inventive world and enable him or her to reach out around the other. Beyond any speaker's bound inventive world lies another: there lie the riches of creation, the great, unbounded possible universe of invention. The knowledge we have is formed out of the plenitude of creation, which is all before us, but must be sought again and again through the cycling process of rhetoric, closing to speak, opening again to invent again. In an unlimited universe of meaning, we can never foreclose on interpretation and argument. ("Argument" 29)

However abundant, the promises of invention are difficult to actualize, according to Corder, for "[w]hen rhetorics come together seeking to occupy the same space at the same time, the encounter can be as sweet and gentle as two people making love, each turning this way and that, wanting to give, not to take, or as hard and hurtful as two tribes making war" ("From Rhetoric" 99). They are difficult even for one person, he says, since "Every soul carries competing rhetorics. I am often if not always at war. . . . I send conflicting messages to myself and to others" (*Yonder* 165).

If inner peace is this elusive, how can we hope to achieve something as implausible as world peace? And how can the discomposed self—who occupies multiple moral and rhetorical spaces—enact this process? It's a question that can't be answered through slogans like "Make love not war." And indeed Corder's point is not simply to lament our apparent inability to forgive or agree, or to suggest that we should turn the other cheek in the face of conflict: "I've not lost my head altogether," he writes; "some conflicts will not be resolved in time and love" ("Argument" 27). His point, rather, is that "if we learn[ed] to love before we disagree[d]," we would still disagree about many things, but we might argue differently: not as a performance or war of words, but as a way of "inviting the other to enter a world that [we try] to make commodious" ("Argument" 26). It's a difference that manifests itself in our roles as arguers and our relationships to adversaries; more specifically, it binds us together as witnesses to each other in the universe of obligation. Holocaust scholars Dori Laub and Marjorie Allard affirm that the proper response to conflict and anguish is not to erase them or wish them away but "to reinvent the responsive other through testimony so as to reconstitute the self as *one who is heard*" (808). While "[a] physical home can never be a home again" under these circumstances, they write, "the place can be reclaimed" (811-12) and truths devised through "reparation of the impaired dyad in the imagined or attempted act of communicating" (808).

Jim Corder's experiences cannot be said to have been the same as those who endured the Holocaust. However, his rhetorical vision embodies an intellectual and spiritual affinity with Holocaust survivors that invites us to read his work as witness testimony, composed in response to war and in the face of profound challenges to his identity. Its purpose, I suggest, is to create moral and rhetorical spaces so vast as to accommodate multiple agendas and irreconcilable realities. The diaspora, imperfect and fraught, pregnant with the potential for understanding, represents such a space.

OBLIGATIONS OF AUDIENCES

I don't intend for this discussion to be perceived as a totalizing narrative about men and war, or even rhetoricians and war. Certainly, a close call is not the only kind of experience that can motivate one to imagine peace, nor does it invariably engender this attitude: For some, it might in fact have the opposite effect, that is, lead them to develop romantic and uninformed perspectives justifying the necessity of war and downplaying its brutality, or equally disturbing, to create symbolic ways to experience combat, for instance, through Civil War reenactments, bullying, hunting, or agonistic discourses—what Walter Ong has called "fighting for life." Consider, for example, statistical evidence that when war ends, violence at home rises (Boyd 4A). Consider, too, convicted Oklahoma City bomber Timothy McVeigh's comment that "Fighting in the Gulf War left [him] angry and disillusioned . . . and the clashes at Waco and Ruby Ridge showed the federal government will use violence 'as an option all the time'" ("McVeigh" 2A). These scenarios perhaps confirm the obvious: that "official," government-sanctioned violence can influence a person to respond violently rather than peacefully. It is, in short, difficult, if not impossible, to generalize about why some people respond with violence to the threat of violence and others respond with efforts to make peace and create understanding.

But if we accept that our own war-consciousness determines our response to war, how might those of us who have not experienced war— or even, perhaps, the "microviolences" described so eloquently by Blitz and Hurlbert—learn from our professional predecessors about how to compose paths to peace? The key, I believe, lies in reading their rhetorical theories as witness testimonies. Langer argues that testimonies can bridge the chasm that separates nonwitnesses from witnesses (including nonparticipants) and survivors. Although many Holocaust survivors declare in their testimonies that "To understand, you have to go through it," Langer maintains that "the sympathetic power of the imagination" can allow those of us who have not "gone through it" to sympathetically enter into the discourse of war and learn its lessons (xv). In listening to witness testimonies, he says, we must "acknowledge our active role as audience to the content of these testimonies" (201) and forge "'new' identities as secondary witnesses" who take responsibility for their knowledge by acting on behalf of others who cannot (195).

The imperative to occupy simultaneously the positions of audience and witness is significant rhetorically for at least two reasons. For one, audiences do not exist apart from writers; indeed, as many theorists of audience have observed, writers are in fact members of their audiences—rereading and revising texts according to their intent—and thus obligated ethically (if not rhetorically) to play by the same rules of behavior that they call for in their own discourses. For another, the concept of praxis demands that we constantly negotiate multiple positions for the purpose of taking informed action; to enact this dialectical process in the presence of others whose very selfhood depends on our responsive action gives substance to what might otherwise be regarded as a low-stakes theoretical exercise. While sophisticated understandings of audience and praxis are not new, when conceived within the context of war testimonies, they contribute an urgency to the goal of transcending the effectiveness model of communication—whose purpose is often to win or be right—and moving instead toward one whose object is to sustain discourse in the hope of deferring war, achieving understanding, and inventing new ways to coexist.

Peacemaking, then, requires reinventing ourselves as both witness and audience, and reinventing our adversaries in the same terms. It requires entering the universe of obligation rather than conniving in its diminishment. When humane uses of language are at the heart—not just metaphorically, but literally, physically at the heart—of these actions, we might move beyond violence and trauma in the direction of peace, becoming in the process "emissaries or legatees of love" (Langer xv).

NOTES

[1]Langer clarifies this point: Nonparticipant witnesses are not the same as survivor witnesses, and it is inappropriate to hold survivors to a standard of action we might reasonably impose on others, since their circumstances often robbed them of their sense of agency. Indeed, says Langer, many are "trying to come to terms with memories of the need to act and the simultaneous inability to do so that continue to haunt [them] today" (183). For the same reasons, he suggests, it is also inappropriate to interpret survivors' testimonies "within the familiar frame of guilt and responsibility" (185). Thus my observations about "relief, guilt, and obligation" should be applied only to nonparticipants.

²In fairness, I should note that Wayne Booth himself isn't so sure that his close call contributed to his interest in what he terms "dialogue-vs.-war." Although he concedes the possibility that "my horror of war underlay all that," he writes:

> For me the influence of coping with my own rival beliefs was much stronger. Raised as a devout Mormon and then encountering doubts of many kinds—and quarrels of many other kinds—I became more and more interested in how to probe beneath surface differences to find common ground. Then the "Chicago school," particularly Richard McKeon, pulled me much further into that kind of inquiry. (Personal correspondence)

WORKS CITED

Barnett, Victoria J. *Bystanders: Conscience and Complicity During the Holocaust*. Westport, CT: Greenwood, 1999.

Berthoff, Ann E. "Index of Richards' Speculative Instruments." *Richards on Rhetoric: I. A. Richards, Selected Essays 1929-1974*. Ed. Ann E. Berthoff. New York: Oxford, 1991. 276-87.

Blitz, Michael, and C. Mark Hurlbert. *Letters for the Living: Teaching Writing in a Violent Age*. Urbana, IL: NCTE, 1998.

Booth, Wayne. *For the Love of It: Amateuring and Its Rivals*. Chicago: U of Chicago P, 1999.

—. *Modern Dogma and the Rhetoric of Assent*. Chicago: U of Chicago P, 1974.

—. Personal correspondence. 7 July 1999.

Boyd, L. M. "Just So You'll Know." *Wilmington Morning Star* 21 February 2000: 4A.

Corder, Jim W. "Argument as Emergence, Rhetoric as Love." *Rhetoric Review* 4 (Sept. 1985): 16-32.

—. "From Rhetoric Into Other Studies." *Defining the New Rhetorics*. Ed. Theresa Enos and Stuart C. Brown. Newbury Park, CA: Sage, 1993. 95-105.

—. "World War II on Cleckler Street." *Collective Heart: Texans in World War II*. Ed. Joyce Gibson Roach. Austin, TX: Eakin, 1996. 18-28.

—. *Yonder: Life on the Far Side of Change*. Athens: U of Georgia P, 1992.

Corder, Roberta. Personal correspondence. 29 March 2000.

—. Personal correspondence. 30 March 2000.

Fein, Helen. *Accounting for Genocide: National Responses and Jewish Victimization During the Holocaust*. New York: Free, 1979.

Fine, Ellen S. "Transmission of Memory: The Post-Holocaust Generation in the Diaspora." *Breaking Crystal: Writing and Memory after Auschwitz*. Ed. Efraim Sicher. Urbana: U of Illinois P, 1998. 185-200.

Freire, Paulo. *Pedagogy of the Oppressed*. New York: Continuum, 1970.

Gurewitsch, Brana, ed. *Mothers, Sisters, Resisters: Oral Histories of Women Who Survived the Holocaust.* Tuscaloosa: U of Alabama P, 1998.

Horowitz, Sara R. "Auto/Biography and Fiction after Auschwitz: Probing the Boundaries of Second-Generation Aesthetics." *Breaking Crystal: Writing and Memory after Auschwitz.* Ed. Efraim Sicher. Urbana: U of Illinois P, 1998. 276-94.

Langer, Lawrence L. *Holocaust Testimonies: The Ruins of Memory.* New Haven: Yale UP, 1991.

Laub, Dori, with Marjorie Allard. "History, Memory, and Truth: Defining the Place of the Survivor." *The Holocaust and History: The Known, the Unknown, the Disputed, and the Reexamined.* Ed. Michael Berenbaum and Abraham J. Peck. Bloomington: Indiana UP, 1998. 799-812.

"McVeigh Blames Government for his Anger." *Wilmington Morning Star* 13 March 2000: 2A.

Ong, Walter J. *Fighting for Life: Contest, Sexuality, and Consciousness.* Ithaca, NY: Cornell UP, 1981.

O'Reilley, Mary Rose. *The Peaceable Classroom.* Portsmouth, NH: Boynton/Cook, 1993.

Recchio, Thomas E. "On the Critical Necessity of 'Essaying.'" *Taking Stock: The Writing Process Movement in the 90s.* Ed. Lad Tobin and Thomas Newkirk. Portsmouth, NH: Boynton/Cook, 1994. 219-35.

Ruddick, Lisa. "The Near Enemy of the Humanities is Professionalism." *The Chronicle of Higher Education* 23 November 2001: B7-B9.

Sicher, Efraim, ed. *Breaking Crystal: Writing and Memory after Auschwitz.* Urbana: U of Illinois P, 1998.

Tobin, Lad. "Reading and Writing About Death, Disease, Dysfunction; or, How I've Spent My Summer Vacations." *Narration as Knowledge: Tales of the Teaching Life.* Ed. Joseph F. Trimmer. Portsmouth, NH: Boynton/Cook, 1997. 71-83.

IV

Theoretical, Pedagogical, and Institutional Issues

12
Bringing Over Yonder Over Here: A Personal Look at Expressivist Rhetoric As Ideological Action

Tilly Warnock
University of Arizona

If this is a memoir or an autobiography, I'll be surprised. Some historian, say, and my mother, were she still here, might testify that I have no occasion to write a memoir or autobiography; the historian, perhaps because I am not notable and have not been where the momentous occurs, my mother, perhaps, because if I've called attention to myself, I ought to be ashamed and to hush. I don't know how to respond to what my mother might have said, but the historian would be wrong: every one of us occasions a book—or a library—because each of us is unique, and because each of us is not unique.

—Jim W. Corder (*Yonder*)

🖝Jim Corder opens *Yonder: Life on the Far Side of Change* by saying that he would be surprised if his collection counts as a memoir or autobiography, yet he continues to justify his collection of personal essays as memoir and as history where "the momentous" happens in everyday life. At the same time, he acknowledges what his mother taught him—not to call attention to himself. Corder has it both ways: He does what he wants in writing but admits he knows better; he mixes and writes across genres but demonstrates his knowledge of genre expectations; he calls attention to himself but admits guilt; and he understands his personal experiences as cultural and political. He may be rebellious, but his mother raised him right.

These are just some of the creative and perhaps duplicitous ways Corder creates an engaging, nonthreatening, even loving ethos in this collection and in his other books and articles. These are some of the ways he identifies with his readers and gets his readers to identify with him. But Corder does far more: He uses personal writing to create identification with readers in order to motivate and move us to further collective action. In doing so he challenges several current assumptions in rhetoric and composition that need further examination and revision if we are to retain and develop further one of our most powerful resources for ideological action.

Limited views of expressivist rhetorics, including personal essays, literary nonfiction, and creative writing, result in impoverished understandings of writing both in the academy and beyond. While such writing cannot be classified as academic discourse in the strictest sense of that phrase, it has already become more persuasive across disciplines and will become more so as university scholars and researchers recognize that the public constitutes one of our major audiences. Scholars in rhetoric and composition who have either denied the ideological power of expressivist discourse or recognized only an individualistic and liberal ideology in personal writing were doing so in part to professionalize the field by promoting traditional definitions of disciplines and forms of academic writing. At this point such disciplining defeats our own purposes.

More specifically, critiques of expressivism were strategies to replace one kind of writing with another because expressivist discourse was understood inherently to promote a certain ideology. Scholars were enacting a traditional argumentative method in academic discourse, denigrating what has gone before in order to make room for what is presented as new. One contradiction or irony in this is that these arguments advocated the familiar replacement model for marketing and consuming goods, instead of a rhetorical model of persuasion that recommends maintaining what has been effective, including what might work in different circumstances, and learning to figure out which among all of the available means of persuasion will most likely work in a given context.

By continuing to deny the values of expressivist writing, we are undermining the grounds of rhetoric, writing, and teaching as actions by people within contexts of situations and cultures. We are often revising ideology and ideological action into subjects to study and teach rather than practice. And because we are not practicing in writing what we are teaching about writing, we undermine the credibility of our own ethos and the ethical positions we value. In addition, as we fail to enact the theories we promote, such as the social construction of meaning, even our own meanings, we act as if our claims are true and not rhetorical. This last point is tricky, of course, because assertions of truth, authority, and verifiable evidence are in fact persuasive today. But it is critical to remember, as we adopt rhetorics of authority and dominance, that all our claims are rhetorical, debatable, and uncertain.

Jim Corder continues to offer us a corrective to our tendencies to promote new theories as truths applicable everywhere and as evidence that what went before was wrong and therefore no longer useful. He advocates both in what and how he writes what Kenneth Burke also advocates, a comic corrective to hierarchical tendencies in language and in symbol-using animals to become "rotten with perfection," to paint ourselves and others into the corner, and to try to make something "crystal-clear" rather than acknowledge we are "on the track of something" (*Attitudes* 86). Through his personal, cultural, creative, critical, and ideological discourse, Corder teaches us to live with the messiness, uncertainties, and ambiguities of life in order to act for change and social justice. He teaches us to mix metaphors, blur genres, tell stories in a nonlinear and nonlogical way, all as ways to help writers and readers identify with each other in order to take further collective action. He teaches us to live with the complexities of life without denying them and to make sound decisions about our choices of language.

Again Burke provides a scene and motive for what he and Corder do with words:

> Hence instead of considering it our task to "dispose of" any ambiguity by merely disclosing the fact that it is an ambiguity, we rather consider it our task to study and clarify the *resources* of ambiguity. For in the course of this work, we shall deal with many kinds of *transformation*—and it is in the areas of ambiguity that transformation takes place; in fact, without such areas, transformation would be impossible. (*Grammar* xix)

It is this attitude toward language and living that allows us to recognize that we should use all there is to use in order to do the work we value, realizing this attitude doesn't mean we have freedom to do what we want if our aim is to convince. As Burke states, the "main ideal of criticism" is "to use all that is there to use" (*Philosophy* 23), but our choices must stand the "'collective revelation' of testing and discussion" (4).

In the following I want to explore further my assertion that expressivist rhetorics can create identification that leads to ideological action; I want also to enact in writing as best I can what I interpret Corder doing in and with his writing. More specifically, I focus in Part 1, "Motives and Consequences of Scapegoating Expressivists," on how Corder challenges and revises conventions of the academic essay, such as a sustained authoritative ethos, verifiable evidence rather than experiential evidence, and air-tight linear and logical arguments. In Part 2, "A Personal Story about Racial Politics: An Indirect Route to the 'Far Side of Change,'" I accept Corder's permission to write autobiographically and historically and turn to my own experiments in writing fiction and nonfiction that led me to examine how the life and murder of civil rights activist Dr. Thomas H. Brewer revised my life. In Part 3, "'We Are Always in a Rhetoric,' If We Choose To Be: Ideological Uses of Expressivist Rhetorics for Ideological Actions," I conclude with Wendy Bishop's critical yet personal essay, "Places to Stand: The Reflective Writer-Teacher-Writer in Composition." She gathers together teachers and writers in rhetoric and composition who advocate and enact expressivist discourse as ideological action; she tells a story that encourages us to reflect on previous decisions we have made about expressivism and to revise them.

THE MOTIVES AND CONSEQUENCES OF SCAPEGOATING EXPRESSIVISTS

Some of the complexities of Corder's use of expressivist rhetorics for ideological purposes become clearer on the first page of the "Preface" of *Yonder*. He explains that he believes "[e]veryone of us occasions a book— or library—because each of us is unique, and because each of us is not unique." He admits the truth of what in the opening sentence surprised him. Characteristically, he wants it both ways:

> But I don't much think that makes this a memoir or autobiography. I do commence and end with myself, and in between I check regularly with myself to be sure that I'm still there. I believe in the significance of regular time and ordinary life, and mine is the one I know, or mostly know. But after all, I don't think that I am subject, object, and occasion for what follows. In that, I'd generally agree with what my mother might have said. (ix)

Corder presents his dialogic selves, an "I" checking with a "myself." His "I," "myself," double-voicedness, and self-reflexiveness can all be read as purposeful constructions aimed to engage and convince readers. He understands his autobiography as cultural writing, at the same time as he respects the "significance of regular time and ordinary life." This desire to have it both ways may also reflect his recognition that as language users we've always got it not only both ways but many other ways. We can emphasize the "I" or the "we," individual voices or cultural voices, extraordinary or ordinary life, in order to do things with words. Without denying the realities of ordinary living, he recognizes the realities of what language constructs and how these constructions may in turn affect ordinary life. Corder is a rhetorician who understands language as rhetorical and revisionary rather than as referential and true. He selects among terms and concepts, such as individuality, life, time, history, and memoir, for specific purposes in particular contexts. He chooses to exercise agency and responsibility, as he acknowledges limitations on his abilities to do so.

In the first essay in the collection, "Restaurant Nora," Corder creates himself (and his reader) as a respectful and traditional academic who, however, writes for "my self's sake" (3). Again, his emphasis on self, on himself as subject, object, and scene, and on the dialogue between "I" and "myself," can be interpreted as a valuing of the individual self. But we can also interpret his ways of writing, particularly how he positions himself, as rhetoric, as how he identifies with and moves readers to do the cultural work to which he is committed. We can also recognize that, for Corder, he is both individual and cultural. The seriousness of his commitment to cultural work is indicated by the fact that he uses his own mother for similar rhetorical and ideological purposes.

In general, then, Corder gives himself and his readers permission to write expressivist discourse with the understanding that such writing—"a scholarly sort of work written in a personal sort of way" (x)—doesn't fit traditional expectations of academic writing. Writing that helps us "see what life is like on the far side of change" extends rhetoric beyond its disciplinary and school borders into public, political spaces. Writing that helps us "see what there is in my circumstances that would let me lift the personal up to public scrutiny, bring the public in to private scrutiny" (x), makes clear that these distinctions, between the private and public and the academic and nonacademic, are rhetorical and revisable constructions that nevertheless have serious consequences in the world. His writing encourages us to recognize the constraints of our own disciplinary theories, terms, and attitudes, in order to figure out what might help us see beyond them.

By admitting that his claims are rhetorical rather than right, Corder invites readers to participate with him in making meaning. He creates an ethos not of personal authority but an ethos of collaborative authority. He chooses to exploit the spaciousness of rhetoric rather than limit himself to logical arguments and evidence. By doing so he gives us license to use our multiple selves and our individual identities, to write about our mothers as well as more public-canonized texts, to waffle and to speak directly and indirectly, and to write across and against the conventions of any particular kind of discourse. He teaches us that we have to be willing to listen, learn, and change; we cannot rely solely on attitudes of certainty and logical proofs to persuade.

Why have we in rhetoric and composition so quickly and fully taken on the ways of the academy, so that we use the same ideas, authorities, and phrases, such as "Foucault" and "discursive formation," even as we advocate resistance and change? Why do we ignore historical circumstances, such as open admissions, in our criticism of earlier teachers and writers? Why would James Berlin, for example, create a taxonomy of writing that locates social-epistemic rhetoric as best, not just for here and now but for all times, schools, students, and teachers? Why is it that rhetoricians seek right answers rather than deal in doubt and uncertainty? Why have we limited the resources of rhetoric? My answer to each question, as I have indicated, is that we in composition and rhetoric have been dedicated to the development and maintenance of our own professional status, and for good reasons. But we have focused on this purpose by focusing off others that are perhaps more valuable in the long run. We have failed to remember the rhetoric of our professional motives and often treat our claims and metaphors as truths. We ignore the negative consequences of our professional goal.

All of this is understandable. What is not understandable, now that we can see in hindsight the limitations of our scapegoating of so-called expressivists and expressivism, is why we do not acknowledge that the claim—that expressivist discourse is not ideological—is rhetorical rather than true. While particular kinds of discourse often do serve specific ideologies in given contexts, all forms become formulas and formulaic over time, and they may be used to serve different ideologies in different situations.

Kenneth Burke again helps me make this point with his recognition that identification between writers and readers is necessarily prior to persuading people to other collective actions. He defines his task as helping people get along with each other in his 1955 "Introduction" to *Attitudes Toward History*. His explanation anticipates what he calls "identification" later in *A Rhetoric of Motives*: "It operates on the miso-philanthropic assumption that getting along with people is one devil of a difficult task, but that, in the last analysis, we should all want to get along with people (and do want to)" (*Attitudes,* "Introduction").

I am also here echoing Burke's recommendations in "Revolutionary Symbolism in America," given at the 1935 first American Writers Congress, the purpose of which, according to Frank Lentricchia, was "to extend the reach of the John Reed Club by providing the basis for a much broader organization of American writers" (21). Given the purpose of the conference, Burke recommends the use of the more inclusive word *people* rather than the traditional Marxist term *worker* because Americans in the 1930s don't want to see themselves as workers and can't identify with the term:

> There are few people who really want to work, let us say, as a human cog in an automobile factory, or as gatherers of vegetables on a big truck farm. Such rigorous ways of life enlist our sympathies, but not our ambitions. Our ideal is as far as possible to eliminate such kinds of work, or to reduce its strenuousness to a minimum. (Qtd. in Lentricchia 27)

Burke's rhetoric here failed, as members of his immediate audience were more aligned with their traditional symbols than with the goal of expanding the organization by identifying with the broader public to convince them to join. For Burke identification is necessarily prior to persuasion. It is a form of collective action, in that both writer and reader must move or change in order to identify and in that it prepares writers and readers for further actions.

Corder uses expressivist rhetorics, particularly personal and recursive narratives, to engage his readers for ideological purposes. As he waffles humorously about his personal essays being and not being autobiography, memoir, history, rhetoric, and story, he clears room for himself and for his readers to use any means available to persuade people to particular ideological attitudes and actions. Burke and Corder are dated, as are many of those identified as expressivists by others. And the goal of getting along with others is certainly not dominant today in the streets, in the media, in classrooms, in homes, or in public spaces. But for this very reason, it is critical that a rhetoric of identification rather than violence and war is available as a means of persuasion.

A PERSONAL STORY ABOUT RACIAL POLITICS: AN INDIRECT ROUTE TO THE "FAR SIDE OF CHANGE"

I want to tell a story about how personal writing moved me "to the far side of change." I could say it all started in a schoolroom in Gillette, Wyoming, where John Warnock and I were teaching a summer Wyoming Writing Project Institute in the early 1980s. I was encouraging teachers to experiment with genres they'd never attempted before, to use words they'd never written, to become characters unlike themselves—to do things with words, with themselves, and with others. My argument was that people learn not only by practicing the same kind of writing over and over but also by trying various kinds of writing. We learn by working with the contradictions, problems, limitations, and strengths that emerge as we experiment with various forms and genres. School too often teaches students by repetition, with the command that students do what they're supposed to do, while many students and teachers learn by experimenting with and contrasting rhetorics and doing what we're not supposed to do. When I saw the writing project teachers using language creatively and having a good time doing it, I realized I hadn't been following my own advice: I told myself I would write a novel before I die.

I could say it all started earlier when I wrote stories in big tablets as a child or when I became an English major in college or when I began graduate school in rhetoric and realized I'd always been interested in rhetoric without knowing what it was. I can also say it all started as I grew up Southern, which means for me listening for how the stated and the unstated contradict and play off against each other.

Fears of Sticking My Nose into Other People's Business

Whatever counts as a beginning, I did write short stories I made up during the rest of the summer institute. And after several months, I began a novel without even thinking. I wrote a short scene with a mother, daughter, and older woman, huddled together in a hallway, talking and occasionally glancing at another older woman curled away from them in a twin bed in the adjacent room. From the beginning the young girl of eleven was Allie, the forty-year-old mother was Caroline, the older woman in her early seventies in the hallway was Mattie, and the grandmother in a coma was always Louise. I somehow knew I wanted to write about how Louise affected the interrelationships among three generations of women and how, even in her death on June 16, 1952, she influences them to shift their relationships to her and to each other in order to survive. I wanted to show how orders and relationships support, suffocate, and change.

I set the novel in June in 1952, the year I was ten. I thought I knew what I wanted to say. The hallway scene grounded me for years as I learned about writing fiction and learned finally why I wanted to write:

> *I feel squashed between Mama and Mattie in the hallway outside Grandma's bedroom. I smell the drawer liner through the cutout flowers on the linen closet door. Inside are stacks of folded, ironed sheets with scalloped edges and monograms tucked deep inside.*
>
> *Mama looks at Mattie, "Grandma will be just fine, Allie. Won't she, Mattie?"*
>
> *Mattie puts her hands on her hips, with her mouth in what Grandma calls a pencil line of despair. She looks toward Grandma who is curled like a baby on her bed. I know better than to say what Mama says to me, "It's not right to tell a story." Mama talks in front of other people like things are hunky-dory, even when they aren't. She says she's being considerate.*

I knew I was drawing on aspects of my own life. Louise's house was the home of a friend's grandmother. Allie lives with her parents, Caroline and Matthew, in the home of another friend. I knew Ida whose son returned home from Korea with frostbitten feet, but I never met him, heard stories about him, or asked about him. Louise dies of liver and colon cancer, as did my mother, and Louise is like my mother in other ways. I am also like my mother in ways and like Allie, Caroline, and Lynn in others. Louise's buddies—Mattie, Sue Marie, Augusta, and Bea—grew from my memories of my aunts and their friends who played Samba on Sunday nights as they shared whiskey and stories about the town. I don't remember a linen closet door with cutout flowers, but I remember that scented drawer liner became the rage in the 1950s.

I finally admitted that the world of the novel was too polite, precious, and intact: It was nothing like my life despite the autobiographical connections. I had the characters under my thumb, but they were pushing hard against my control, as the goldfish in Allie's head crash against her insides, forcing her to speak the thoughts and feelings she has learned to silence, in a dialect she has known all of her life but known never to speak aloud. One day, out of the blue, the Chattahoochee River flooded the fiction, washing in all sorts of characters and destroying state lines and other arbitrary borders that powerfully separated people by race, class, religions, and genders. Phenix City, the "Sin City" across the river in Alabama, merged with the fictional Riverside. Allie faced directly the mill workers and soldiers from Ft. Benning. Black characters, who had been only servants in the homes of whites, began to speak out and make room for their own stories and lives.

I constructed the novel from bits and pieces of memory, imagination, dreams, stories, lies, gossip, and lessons that I had to shape in order to make sense, as Allie has to weave together the fragments she learns about from one-sided telephone conversations she overhears; from stories people tell at home, in church, in school; from letters and photographs; from fiction; and from newspapers. As her parents talk on the morning of June 16, 1952, about the murder the night before of Dr. Phillips and Ida's son, Thomas, a vague memory took form. I told myself I was writing fiction and didn't need to do research to find out if a black doctor had been murdered in Columbus when I was young. But as Allie questioned her life, I began doing research on what I had denied all of my life.

I first called the Columbus Public Library to get newspapers published between June 15 and 17, 1952; I told myself I was only checking the weather. When the package arrived, I combed the newspapers for ads, movies, and announcements, ignoring at first the news about local, regional, and national civil-rights actions. The broadest context for the fiction from the beginning was the Korean War and the fact I remembered that it was the first war where soldiers of all races lived as well as fought together. I finally learned why I was writing: I wanted to find out about the murder from which I shielded myself and from which others had protected me. I wanted to learn about Dr. Thomas H. Brewer, his life and his death, and about Luico Flowers who was charged but acquitted of his murder in self-defense. I wanted to examine the history of civil rights in Columbus, Georgia, and in my own family and life. I wanted to understand relationships between Dr. Brewer's murder and the fictional Dr. Phillips' murder. I saw the arc of Louise's death on June 16 intersect with the arc of Dr. Phillips' murder the night before as a shift from one order to another. I understood how this intersection sparked the storm in Allie's mind that forces the goldfish in her head to speak in a black dialect she has known since birth; I recognized that in writing I was also speaking a silenced language in me. I figured out how Allie's speaking the goldfish taught me to speak what I had always known but never spoken or written. I wrote fiction that led to nonfiction and history that many understand as fictions and denials of truth.

In trying to understand through writing and research, I tested genres and combinations of forms to figure out how to tell my interpretations. I questioned official versions that Luico Flowers murdered Brewer in self-defense and then committed suicide a year later. I examined how a culture politely and violently educates people into racism and kills friendship between a black man and a white man. I wanted to understand how we can revise our lives.

And Just Who Do You Think You Are?
You Have No Occasion to Write a
Memoir

When I began writing nonfiction, the story of how writing fiction led to
nonfiction and then to more fictions, I called Bill Winn, editor of the
Columbus newspaper, to ask if he knew about a murder when we were
young. He did and said he had written articles on the life and death of
civil rights activist Dr. Thomas H. Brewer, who was murdered in 1957. I
was fourteen, not eleven, at the time. I called the Columbus library again,
asking for copies of files on Dr. Brewer. After reading these, I moved the
first chapter that takes place in the hallway to chapter two and wrote a
new chapter one "Everything is Satisfactual, Like the Song Says":

> *"Who's Dr. Phillips?" I ask, bumping the door open with*
> *my knee.*
> *"Good morning. How's my girl?" Daddy looks up from the*
> *newspaper.*
> *Mama frowns, "Allie, I've told you countless times not to*
> *enter a room asking questions about conversations that you*
> *were not invited to join." She puts a curve of honeydew in*
> *front of me and turns to talk to the refrigerator, "Your father's*
> *having scrambled eggs with cheese and bacon. Do you want*
> *the same?"*
> *"Plain eggs, please, and I'm sorry. But who is Dr.*
> *Phillips?" I look from one to the other. "I wasn't*
> *eavesdropping. I could hear you talking from the top of the*
> *stairs."*
> *"Not somebody you know." With her hip Mama closes the*
> *door to the refrigerator and to Dr. Phillips, whoever he is.*
> *I turn to Daddy reading the* Daily Star, *hoping he'll*
> *answer my question. Mama sighs at the sink. I scoop the soft*
> *lime melon slowly and suck it into my mouth, checking to see*
> *if they notice. The goldfish in my head swim fast into a streak,*
> *crashing against my insides, making my head hurt.*

Allie's opening question, "Who's Dr. Phillips?" was a version of my
question, "Who's Dr. Brewer?", a question I was afraid to formulate and
research. In contrast, Allie at eleven asks the adults around her about Dr.
Phillips, though she never gets a straight answer. She reads the local
newspaper at midday and learns what her parents and others have
hidden from her. She realizes that her world has in part been created by
the silences, deflections, stories, and denials of people she loves and by
her own failure to question and seek answers for herself. As the goldfish
fly from her mouth, Allie falls apart but begins to find new languages,
identities, and ways to survive.

At this point I inserted nonfiction in between chapters of fiction to
tell the story of writing fiction that led to nonfiction. I include short
essays on relationships between autobiography and fiction, biographical
sketches of people associated with the fictional characters, newspapers,

and movie ads. I write about Dr. Brewer's actions to improve the segregated schools for black students, to integrate public facilities, and to increase voter registration. I draw parallels between how Dr. Phillips' murder in the novel affected Allie's life and how Dr. Brewer's murder revised mine. The following introduction explains how expressivist writing can lead to ideological action:

Introduction
I thought I was writing a novel, but the novel rewrote me. I wanted to make something up. What I learned was that something had already made me, something I did not remember, something dark like a shadow and heavy like water. It was something I didn't know consciously and therefore could not remember in words. I could not analyze the "consciousness which is unclear to itself."

Writing let the goldfish speak and taught me what I already knew but was too afraid to speak or write. Many revisions over the years also changed my understanding of what novels can be and do and what can count as fiction, nonfiction, history, politics, stories, lies, and truth. At one point I called what I was writing fiction. I then called it a nonfiction novel that dramatically juxtaposes fact and fiction. I now call this book writing that revised me and my understanding of family, race, class, gender, and culture.

"WE ARE ALWAYS IN A RHETORIC," IF WE CHOOSE TO BE: IDEOLOGICAL USES OF EXPRESSIVIST RHETORICS

Today, we in rhetoric and composition often interpret expressivist rhetorics as writing that betrays the political origins of rhetoric and the cause of rhetoric to fight injustice and inequality. We often interpret this claim as truth not rhetoric, but we can, instead, acknowledge as Corder does that "we are always in rhetoric." We must also, I believe, acknowledge that even this claim is rhetorical—purposeful and consequential—but not true. Like all assertions, this claim and ones such as "rhetoric is ideological" or "rhetoric is not ideological" can be read as assertions of metaphors to explore for what they yield rather than as truths to accept. We must continually assess the gains and losses of our aims and assertions to revise them for new contexts.

Similarly, we can admit that expressivist rhetorics may be more persuasive than contemporary academic discourse in particular situations, for example, with our students at times and with some public audiences. Wendy Bishop, along with other scholar/teacher/writers such as Nancy Sommers, Mike Rose, Richard Miller, Lynn Bloom, and Sondra Perl, challenges attacks on expressivist rhetorics by what and how they write. Bishop, in "Places to Stand: The Reflective Writer-Teacher-Writer in Composition," provides a useful account of current debates about expressivism. She begins with a quotation on page 98 from Corder's "From Rhetoric into Other Studies" that asks how we can deny the ideology of our own discourse as we claim all discourse is ideological:

> We are always in rhetoric. We may see those others in rhetorics not our own; if we do, they are likely to seem whimsical, odd, uninformed, selfish, wrong, mad, even alien. Sometimes, of course, we don't see them at all—they are outside our normality, beyond or beneath notice; they don't occur as humans. Often as not, we don't see our own rhetoric; it is already normality, already truth, already the way to see existence. When we remark, as we have become accustomed to remark, that all discourse is ideological, we probably exclude our own. It is the truth, against which ideological discourses can be detected and measured. (Qtd. in Bishop 9)

Bishop also sets up her "admittedly personal essay" by explaining how social constructionists have defined expressivists for their purposes (12):

> [K]ey-expressivists (so called, not self-labeled) are frequently cast as convenient straw-men, as now-aging, no longer compositionally-hip, and therefore slightly embarrassing advocates of 1960s touchy-feely pedagogy from which professionals in composition are currently trained to distance themselves. This anti-expressivist encouragement takes place despite The Many C's continued publication of works by teachers-who-are-writers; my continuing list of these includes Lynn Bloom, Lillian Bridwell-Bowles, Robert Brooke, Jim Corder, Elizabeth Rankin, Mary Rose O'Reilley, Michael Spooner, Lad Tobin, and Kathleen Blake Yancey. (10)

Her aim is not to defend writer-teachers but

> to examine why it seems so uncanny to me today to be hearing the term post-process (been there, done that?) percolating through the air at conferences, and to find it assumed that expressivists don't do other things ("things" are often represented by the word "theory") because they "can't" not because they choose not to. (11)

She continues her "scholarly sort of work written in a personal sort of way" (x) by examining several recent articles that address issues of expressivist pedagogy and discourse. For example, Bishop uses Fishman and McCarthy's 1992 "Is Expressivism Dead?" to summarize critiques of expressivist pedagogy by Bartholomae, Bizzell, Berlin, and Trimbur:

"These scholars' critiques include the charges that 'expressivists' keep students in a state of naiveté, don't prepare them for the languages of the academy, abandon them to the forces of politics and culture and 'emphasize a type of self-actualization which the outside world would indict as sentimental and dangerous'" (qtd. in Bishop 11).

She adds that Gradin in *Romancing Rhetorics* "points to the way she believes critics have overlooked certain commitments and potentials of expressivist pedagogies, 'contrary to how social epistemic hard-liners would have us believe, the important things [positions] that Berlin outlines here are not solely the province of social-epistemicism'" (qtd. in Bishop 11). Bishop quotes Gradin further to make her point that expressivist pedagogies don't prevent ideological actions of various kinds in the classroom:

> Nothing restrains an expressivist teacher from asking students to examine who gains from their "personal visions," from their "individualistic stances." Examining racism and sexism is easily done in the expressivist classroom. Moreover, critical reflection . . . is a major facet of expressivist theories and practices . . . the expressivist classroom can resist disempowering social influences, use interdisciplinary classroom methods, and posit a social understanding of the self. Expressivist rhetoricians certainly can be self-critical and self-revisionary. (Qtd. in Bishop 11)

She understands antiexpressivism as a result of "rapid professionalism" and the "need to appear ever-more scholarly, historical, and theoretical" (12). She advocates "overcoming our fear of this other" (13) and providing places for the teacher-writer to stand that are not "marginalized places: the citation, the staffroom exchange essay, the poem about classrooms" (24).

In this article and elsewhere, Bishop and many writer-teachers demonstrate along with Corder the rich possibilities of expressivist rhetorics. I want to affirm the practice of using such ideological rhetorics to create identification across differences that doesn't deny differences. To ignore expressivist rhetorics and practice only conventional academic discourse is just one more way to say, "Hush your mouth. Mind your manners, do what you're supposed to do, and don't cause trouble."

WORKS CITED

Bishop, Wendy. "Places to Stand: The Reflective Writer-Teacher-Writer in Composition." *College Composition and Communication* 51 (1999): 9-31.
Burke, Kenneth. *A Grammar of Motives.* Berkeley: U of California P, 1969.
—. *Attitudes Toward History.* Berkeley: U of California P, 1984.

—. *The Philosophy of Literary Form*. Berkeley: U of California P, 1967.

Corder, Jim W. *Yonder: Life on the Far Side of Change*. Athens: U of Georgia P, 1992.

Lentricchia, Frank. *Criticism and Social Change*. Chicago: U of Chicago P, 1983.

13
A More Spacious Model of Writing and Literacy

Peter Elbow
Professor Emeritus, University of Massachusetts

There is a great deal of writing going on in everyday life, writing outside of school, work, and "established" publishing. Yet, very little is known about the scope, nature, use, and social context of writing in everyday life; indeed a good case can be made that writing in everyday life is nearly invisible. . . . it is frequently not even considered to be writing, and the people who produce such writing are not usually considered writers.

—David Bloome ("Introduction")

Life is real, and the artificial compartments we create for it don't work.

—Jim W. Corder ("Hunting")

ಎ Jim Corder was always trying to make more commodious space for the messy reality that artificial compartments fail to account for. As he once said to Keith Miller (in a discussion about Kinneavy), "Taxonomies always leak"—or as Miller sums it up in the same e-mail to me, "He regularly tears up boxes" (April 14, 2000). In this chapter I'm trying to tear up some boxes about the nature of writing—the boxes that are particularly powerful because they are unspoken taxonomies inside the heads of so many people. I will end up with a few boxes of my own—but in fact they aren't so very different from what Corder made when he was in a box-making mood. In particular, I'm trying to get these boxes out on the table where we can see and critique them.

THE CURRENT, DOMINANT, DEFAULT, UNSPACIOUS MODEL OF WRITING

When I ask people about their history as writers, many of them refer only to writing they do in school or on the job. It's as though they don't notice or remember any other writing they've done. Some people even say "I never write"—when in fact they do write. The dominant model of writing in composition and the wider culture doesn't make space for all this other writing that people do—writing that we might sum up (speaking roughly) as "everyday writing."

School

It's not surprising that people associate writing with school. We learn to speak at home as infants and use speech in every facet of our lives, yet most people learn to write only in school. For that reason, the following assumptions are remarkably dominant—even among teachers and scholars:

- Writing is something that teachers make you do.

- In order to write, you must already have mastered the alphabet and the conventions of grammar, sentences, and paragraphs.

- It's difficult to master these foundational skills so well that you don't make mistakes.

- When you write, you give your words to a teacher, someone who has authority over you and who almost always gives you some kind of evaluative response. Even when teachers are very busy, they usually at least circle a few errors—or give some kind of grade (even if only a check—perhaps with plus or minus).

- Because it's so hard to master the foundational skills of writing and because writing is virtually always evaluated, most people experience writing as harder and more dangerous than speaking. Most people feel inadequate and anxious about writing and seldom write unless they have to.

But more and more people also associate writing with the workplace. A flood of recent research has shown how much goes on. Faigley and Miller write that "People in professional and technical occupations . . . on the average write nearly 30% of their total work time" (564). My students no longer complain that they'll never need writing outside of school. "If you really hate writing," I tell my resistant ones, "you're better off becoming an English teacher than an engineer." (See Odell and Goswami for more on workplace writing.)

But the prevalence of workplace writing doesn't much change the complex of assumptions I've just laid out. Like the teacher, the supervisor or boss assigns and judges one's writing from a position of authority. In fact, the weight of authority or risk in workplace writing is usually greater: The writing is often also given to fellow workers, a customer, a client, or members of some other business. Such audiences have even more authority than the boss, since their disapproval means loss of money, business, reputation, a project, or even one's job. So workplace writing is no less difficult, dangerous, or audience directed than school writing. (One big difference: Workplace writing is more often collaborative than what students encounter in most classrooms.)

Of course, I've neglected a third kind of writing that most people notice—but seldom identify with. A few people warrant a special title in our culture, *writer*, because they write seriously by choice. Even when students or workers spend lots of time writing, they seldom think of themselves as writers. Thus the dominant model has another clause: Writing is also something done by special people who have special talent and dedication and do it by choice.

Is there a conflict here? Writing as something that most people avoid and some people seek? Not really—once we add a crucial term that was tacit above: *real* (as in "real writer" and "real writing"). This yields a single fuller version of the default model of writing in our culture:

> *Ordinary people only write when forced to do so (in school and on the job) and what they produce is ordinary writing. Real writers write by choice and what they produce is real writing.*

In fact there's even some question about how much choice real writers have in the default model—so deeply linked are the notions of writing and compulsion. That is, even real writers often put it out that they actually have no choice. Rilke gives the classic formulation: "If one feels one could live without writing, then one shouldn't write at all." Writers seem to like saying this. Anecdotes are commonly told of famous writers fending off would-be writers by asking, "Why do you write?"—and then dismissing them unless they answer, "Because I cannot *not* write." I've had many occasions to notice how deeply offended many people are by the concept of freewriting—offended, that is, by the attempt to link writing and freedom, especially for ordinary people.

So the default model links ordinary people and real writers not just around compulsion but around struggle:

> *Whether you are an ordinary person who produces ordinary writing or a real writer who produces real writing, your process of writing will be difficult and it will involve struggle and even pain.*

The truth is that many real writers get substantial pleasure from writing. But, interestingly, they mostly don't talk about this in public. Indeed, it is widely believed—and writers seldom do much to dispel the illusion—that the very identity of *writer* involves not only struggle but suffering and isolation. Alcoholism is still in fashion for writers—suicide a bit passé. (I know of more than one instance of an online writing class where enjoyment of the online setting confused students about the very concept of writing: "They don't consider what they are doing as writing because they like this writing; they seem like determined nondancers found tapping their feet" [DiMatteo].)

TRYING TO NOTICE EVERYDAY WRITING

School may be the archetypal site of writing, but we in composition tend to complain about how little students are asked to write in school and college. (Applebee gives shocking if not current data about how little teachers ask students to write.) But we need to notice the assumption that our complaining rests on: that everyone ought to write a lot. I believe this wholeheartedly (so I'll probably never stop complaining), but what's important is to notice how new this assumption is. Surely no other society has ever made this assumption—nor gotten virtually all people into school and writing.

It's probably this assumption that leads us not to notice all the writing people do outside of school or workplace. We teachers spend so much time making people write when they don't want to that it's hard for us to see and believe that folks would do it by choice. Our students are so struck with compulsory writing that they can't believe it's writing when they do it by choice. Faigley and Miller are typical when they say, "[In our research on two hundred college-educated professionals,] we found that people do not write much off the job" (562).

"Do not write much." A rubbery phrase. My claim here is that we won't have a good understanding of what writing is until we notice that people do, in fact, write a remarkable amount by choice out of school and workplace. There's a growing body of research to document this everyday literacy (see for example, Barton, *Literacy*; Barton, "Social"; Barton, Bloome, Sheridan, and Street; Barton and Ivanic; Bloome; Duin and Hansen; Finders; Graff; Hamilton, Barton, and Ivanic; Howard; Sheridan, Street, and Bloome; Tebeaux; and also various of the publications of the Centre for Language in Social Life). I won't try to summarize their rich and complicated findings, but let's just look again at Faigley and Miller's "not much." Out of a random sample of college graduates in professional jobs, Faigley and Miller (1982) note:

- 10 percent "wrote for or edited some type of publication off the job (for example, a regional Audubon Society newsletter)";

- 12 percent "kept diaries or journals, but only two of them wrote daily entries";

- "Respondents wrote on the average less than one personal letter a week."

That last sentence is telling. Anything less than *most* of them averaging *more* than one letter a week is simply dismissed as a nonfinding—not worth transmitting as data. In fact, their laconic sentence implies that a lot of their subjects (even if less than half) wrote a letter more than once a week! It sounds as though most of them wrote letters at least every few weeks. And this was decades before the avalanche of e-mail writing.

Let me flesh out the picture of everyday, self-sponsored writing by exploring some of the kinds of writing that go on:

Journal-Writing

Many adolescents start a journal, and plenty keep it up—a few for their whole lives (including 12 percent of Faigley and Miller's professionals). Think of all the attractive "blank books" sold; not all are left blank. My own experience with journal-writing is probably typical of many others: I've only written regularly during a few periods of my life (from a few months to a few years). For the rest of the time, I write only if I am perplexed or upset or doing something interesting that I particularly want to remember (such as being on a vacation).

Serious journal-writing shades over into the extended jotting that so many people do when they are trying to figure something out. Many people who "don't keep a journal" fairly often explore their thinking when they are perplexed about, say, whether to take a job or break a relationship.

Letters

This is a huge genre of writing—with many subcategories. First, there are conventional letters to friends and family. If we think about how many people *used* to write frequent and lengthy letters (and how parents *used* to make their college-age children write home once a week), we are liable, like so many dyspeptic commentators, to think of letter-writing as a "dead art." But if we look with an unjaundiced eye, we have to acknowledge that a remarkable quantity of letters still get written by choice.

- Notice how many people write letters to the editor of local, large city, and even national newspapers—and magazines. Repeated reader research indicates that these are the most heavily read items in the newspaper. (Stotsky gives some remarkable figures on this matter: "By 1976, the Associated Press Managing Editors' [APME] Red Book estimated that letters to the editor column were providing an outlet for nearly 2 million letter writers per year; undoubtedly, many more millions were being sent to editors but never published" [231].)

- Related to this is the huge volume of letters written as a form of lobbying—to elected politicians and appointed officials and also to corporations. (Stotsky notes that by 1984 there were 200 million yearly letters to the House of Representatives, 41.5 million to the senate—a five-fold increase in 12 years. Drawing on research by Buell, Stotsky notes interesting findings from a survey of 163 people who wrote letters to the editor and 2,000 nonletter writers: "[I]ncome and education were not highly significant factors separating writers from nonwriters" [232].)

- Consider even the growing number of Christmas letters that families put out, and the undiminishing stream of complaint letters that people still write.

- We see another interesting and flourishing subgenre of letters in the ubiquitous notes that children pass to each other in elementary, middle, and even high school. Students often compete with elaborate "formatting": complex decoration, flamboyant lettering, and intricate folding. Almost all school children take part in this form of literacy. Teachers sometimes try to stamp out this kind of writing but invariably fail (see Finders on "hidden literacies" in junior high school).

- The emergence of the Internet has fueled a phenomenal growth in e-mail letters. They are so inviting because they are so easy. They border on conversation. In contrast to conventional letters, e-mail correspondence invites brevity, yet also invites frequent reply.

- Not a few people even converse in writing in real-time in chat rooms. There are reports of some addiction to this kind of writing. E-mail and the Internet have even made it possible for people to engage in writing as sexual activity.

- E-mail correspondence shades into participation on e-mail lists and even the construction of home pages. People pay a good deal of time and money to do this kind of writing.

Personal, creative, and nonfiction writing by people who don't aspire to publication or to the identity of writer: I'm struck with how many students, once they get to know and trust me, show me their poems. These poems are often an overflow of spontaneous feeling: love, admiration, happiness, sorrow, awe, anger—and they are often labored over. Sometimes they are written to share, sometimes just for the writer alone and for pleasure of the writing—to help the writer express and remember and preserve. Most serious departments of English can be said to have been trying to stamp out sentimental, sincere, corny, purple poetry—but to no avail. (Barton and Padmore, 1991, speak of nine out of twenty in their sample of very ordinary people writing at least some poetry.) Narrative too. Many people write out accounts of experiences they have had—scrawled fast or written with care—with no expectation of publication. "I want to capture what it was like." "I don't want to forget all these things." "I want to preserve these stories for my family." My students show me that it's not just older people who want to record what has happened to them in their lives. The picture gets richer if I note just two important sites for this kind of amateur writing:

- Anne Gere documents how informal reading and study groups have been a staple of our culture for more than a century.

- Many senior centers have programs where participants write about their lives—sometimes sharing their writing with school-age children who also write.

- Every summer there are something like 175 area writing project sites based on the Bay Area model: gatherings of ten to forty teachers for four to six weeks. A central feature of these projects is lots of self-sponsored writing.

I'm struck with students I meet who write some form or another of cartoons, comic strips, or comic books—sometimes shading over into dungeons-and-dragons kinds of writing. This sort of writing often carries drawing with it. We can see this kind of writing more clearly if we notice how similar it is to the drawing (and music) that so many young people create by choice—from crude sketches to highly talented finished drawings. Often this drawing too is not for publication—just for fun and for showing to friends.

"Not Aimed at Publication"

With that simple phrase, I skate over something slippery. Much of this everyday writing is indeed intended to be shared with a friend or even with a group. More and more people "publish" their writing by putting it somehow on the web or mailing it to more than one friend/reader— sometimes e-mailing it to a circle of sharing friends. When people contribute to an e-mail listserv discussion or construct a home page for the web, they are moving one step closer to conventional publication—yet, importantly, taking publication into their own hands. More and more people move closer still when they harness new technologies and write for or produce a Zine. If people mail out a hundred copies of a Christmas letter, have they published it? All these activities make a mishmash of what used to be a simple distinction between published and unpublished writing.

Writing We Forget to Think of as Writing

Plato warned us in the *Phaedrus* that we'd lose our memories if we fell for the newfangled technology of writing. But we didn't listen to him, and the result is that most people make lists—often daily—and many people spend considerable time and energy at it. Two small instances have always intrigued me: Whenever a child is born, parents almost invariably find themselves making lists of what needs to be done and even of possible names. Whenever someone dies, survivors almost invariably find themselves involved in serious list-making to try to deal with all the things that have to be done. Births and deaths are not uncommon—and writing is a technology widely resorted to when feelings threaten to overwhelm.

Scholars of language acquisition acknowledge that infants are genuinely talking when they babble, but there is too little recognition that tiny children are often writing when they scribble. Indeed, this scribbling is often closer to writing than babbling is to speaking—closer, that is, to meaningful verbal behavior that contains an actual message. Toddlers can often tell you exactly what they are writing and can often "read" it back after a few minutes—or even later (Graves, 1983).

Unpublished Writing by People Who Seek to Publish; Some Consider Themselves "Real Writers," Some Only "Aspiring Writers"

As we stand in a library or bookstore, we may feel overwhelmed at how much is published. But here again we have to look at things the other way around: Think of how much is not published! For every published novel, there are probably fifteen or twenty that are written and rejected. The ratio in poetry is far steeper. In this connection let me point specifically to all the writing done in MFA programs, elective writing courses, and undergraduate majors or concentrations in writing.[1]

Let me quote a different passage from Bloome to sum up my argument that we need to stop thinking of self-sponsored independent writing by ordinary people as the exception and see it instead as central:

> [T]here is a breadth and depth of writing in the general public, among "ordinary" people, that has not yet been revealed or understood by scholarship on writing and literacy. Similar suggestions have been made by historians of writing, such as

> Howard (1991) and Graff (1979) with regard to breadth (how
> many people write and the broad range of uses and genres of
> writing); but as yet there has been little investigation of the
> depth of writing (the importance people place on writing and
> the ways they use it to organize, structure, and give meaning to
> their lives, as well as to make changes in their own lives and in
> their communities and society in general). (7)

I hope this chapter serves as a call for more research.

TOWARD A MORE SPACIOUS MODEL OF WRITING

We can't deny all the actual writing experiences that have given rise to
the dominant default model in so many people's heads. But we do need
to deny that this model tells the whole story of how writing goes in the
world. There are too many contradictory experiences that are just as real,
especially those uncovered by researchers into everyday literacy. Thus
any adequate model for a complex entity will contain contradictory
elements. (When Corder said that taxonomies leak, he also recognized
that they collide and dent each other.) So let me provide some fragments
that contradict the dominant model. In the end I'll suggest an approach
for how we might put all this rich material together into something more
coherent.

Whereas the dominant model emphasizes compulsion, we now have
lots of evidence that, in fact:

> *Writing is something ordinary people do—and by choice.*

Whereas the dominant model emphasizes the need to learn the
conventions of literacy, in fact:

> *In order to write, you don't need special preparation or
> training at all.*

The classic case is that of toddlers who scribble meaningfully and
kindergarteners who write with invented spelling. They write anything
they can say (Graves). There's an instructive analogy here with musical
literacy: the well-known Suzuki method of teaching very small children
to play stringed instruments. We see an important larger principle here:
In all three cases—the scribbler, the kindergarten writer, and the toddler
violinist—a breakthrough comes from short-circuiting the distinction
between orality and literacy. The ear is a prime doorway into literacy.
(People tend to associate "literacy" with reading more than writing—but
the word obviously pertains to "letters," both reception and production
[see my "War"].) Researchers into everyday literacy show adults busily
writing without mastery of the rules.

Whereas the dominant model emphasizes an audience with authority, in fact:

> *A great deal of writing is intended for peers, not for someone in authority.*

Much of the vast letter-writing that occurs on paper and online is for peers—as are multitudes of stories and poems and accounts of experience. When people write for peers, they begin to break out of the commonly internalized sense that the reader of a text always has authority over them.

Whereas we tend to see readers in binary terms as either authorities or peers, in fact:

> *A certain amount of writing goes to an important third kind of reader not much noticed in theoretical models of writing, the ally reader.*

Some feminists have been reminding us that it is not so rare to read as a person who cares for or even loves the writer (Noddings; Schweickart). When caring friends share independent writing with each other, the activity is often more about friendship and closeness than about writing—even when writing is a proximate goal. Even teachers occasionally slide into the role of ally reader when students hand in writing that is a cry of pain or a call for help. When teachers respond to this writing, they usually bypass the question of how to help the writing; instead they try to figure out how to help the person. In fact, by the time a semester is half over, teachers often find themselves genuinely caring for many of their students. (This is one more reason why many find grading so difficult. The converse is worth noting: Most of us find it easier to be an ally reader of a piece when we don't have to grade it.)

Whereas the dominant model emphasizes that writing is virtually always for an audience, in fact:

> *A great deal of writing is private or for the writer alone.*

It's not just all the journal-writing people do and all the lists and jottings and sketches that people make—most of which remain private. In a real sense, much of the writing that *most* thoughtful writers produce is private: never seen by anyone else. If this sounds odd, it is only necessary to remind ourselves that few people write anything and just hand it to readers as it is (especially since the coming of computers). Most people read back over what they've written and almost always make at least a few small changes—often many changes. That first draft (or second or fifth draft) was literally private.

Someone might object that such writing is not really private— because it was intended for other readers. Fair enough. Yet this very mental condition—thinking about readers as one writes—bears closer examination. In a real sense, people write their early drafts as though readers might break down the door and read their unrevised writing. When people finally realize that all this early writing is genuinely private, they tend to learn to be more relaxed, adventuresome, and productive in their drafting process. Here is Chomsky on this point:

> I can be using language in the strictest sense with no intention of communicating. . . . As a graduate student, I spent two years writing a lengthy manuscript, assuming throughout that it would never be published or read by anyone. I meant everything I wrote, intending nothing as to what anyone would [understand], in fact taking it for granted that there would be no audience. . . . [C]ommunication is only one function of language, and by no means an essential one. (Qtd. in Feldman 5-6. For more about theoretical conceptions of private writing, see Elbow, "Defense.")

The present emphasis on social-constructionist theory sometimes obscures the important role of symbolic activity that is kept private. David Burrows points out a simple but profound fact: "The greater part by far of the world's music is produced by solitary hummers and whistlers [and singers] for no one's consumption but their own" (34). The same probably goes for drawing, sculpting, and perhaps even dancing.

Whereas the dominant model reflects most people's experience that speaking is safer than writing, in fact:

Writing is usually safer than speaking.

The misunderstanding stems from habits of use in our culture: We usually learn to write in school where our writing is always critically evaluated by someone with authority, and we usually use writing for more serious or high-stakes purposes ("putting it down in black and white"). In contrast, we characteristically use speech in many low-stakes, chatty ways. But the inherent possibilities in writing and speaking point in the opposite direction. Writing is actually the ideal medium for exploring thoughts or feelings for no one else's eyes or ears. Speaking is almost always for a listener (even if we occasionally talk to ourselves).

Whereas the dominant model assumes that writing always gets some kind of response, in fact:

Writing is often shared with a reader who gives no response at all.

We see this most obviously in the following situations: personal letters that get no reply, published letters to the editor that get no response, lobbying letters to officials (though there is sometimes a pro forma machine reply), and most contributions to online lists. But let me explain something less obvious but more important.

Like most people, I never used to imagine a model of writing with no response. Then one evening I was sitting in a tiny auditorium listening to a few poets and fiction writers read from their work. It finally dawned on me: This is a no-response situation. Of course you could insist that there is always a bit of response (applause, yawning), but we mustn't let a small overlap blind us to a large distinction. The writers were not asking for any response other than the ritual of applause. They would have been insulted if someone raised their hand and tried to give feedback. These people were writers in a crucially neglected sense of the term: People who assume that response to their text is something they get only when they ask for it. And they knew the value of not asking for it. In that room the writers and readers had tacitly struck a bargain: Writers get to read; audience gets to listen and enjoy.

Most people assume it's anomalous when writing gets no response at all, but that just illustrates the confusion caused by the dominant model. In truth, no response is the fate of most writing in the world. When people write books, stories, poems, newspaper articles, and memos, the words go out and that's pretty much it. Indeed, the writer is lucky if readers *read* the words. Only very occasionally do we get something back in our mailbox. Many books are reviewed, but many are not—perhaps the majority. Magazine articles and newspaper stories are virtually never reviewed. Success with a memo usually means getting no response at all. When most people share what they've written with friends, the response is most likely to be appreciative (and nervous) thanks rather than any sustained response.

It's crucial to notice, then, that even though most people write with the *expectation* of response—usually with the expectation of criticism— nevertheless, the norm for most texts in the world is no response. Of course we may *request* response and criticism from friends and colleagues—and it is often helpful to do so. And of course it's hard to reach many readers without an evaluative Yes or No from an editor or supervisor. Yet it's amazing how often these gateway figures say only Yes or No—and give no other response. Most people's relationship to writing would change for the better if they adopted a more accurate model of writing in the world and understood that getting no response is or can be as common as getting a response.

Whereas the dominant model assumes that response is invariably judgmental or evaluative, in fact:

> *When writing gets a response from readers, that response is*
> *often nonevaluative, nonjudgmental, noncritical.*

The dominant model is preoccupied with judgment because it is preoccupied with school and workplace writing. If I ask anyone to respond to a piece of my writing, they will invariably feel they have to say something about how good or bad it is, and to give me advice for making it better. Many people cannot even respond to their own writing except by evaluation and criticism. Through a process of internalization, they have created a critical reader-in-the-head who is always looking for faults.

Letters are the most obvious case of writing that often gets a response without evaluation. Yes, respondents often say how much they enjoyed our letter, and occasionally how nicely it was written. But if someone replies to our letter by evaluating its strengths and weaknesses, we know immediately that something has gone awry in the communicative transaction. We see the same thing with e-mail letters, notes passed between students at school, fan letters, and many contributions to lists or the Internet: There may be a response but no evaluation. When people write poems or stories or accounts of experience for friends and loved ones, yes, the recipients often give an ostensibly evaluative response: "I really love your wonderful poem!" But actually, this response is more appreciation of a gift than evaluation of writing. (For more about nonevaluative responses to writing, even in a school setting, see Elbow "Taking Time Out" and Elbow and Belanoff.)

A LARGER THEORETICAL MODEL FOR WRITING

All these fragments, then, contradict the dominant default model and the experiences that gave rise to it. Any theory of writing needs to be complex, and in particular to account for the contradictory behaviors of people and the differing conditions and experiences in which they write. But I can suggest a schematic diagram that encompasses at least most of these complexities. My visual model concentrates on audience and response. The key here is to recognize that audience and response are independent variables. That is, even though many people associate certain audiences with certain responses—for example, teachers with evaluation and the self with noncriticism—this is simplistic. Teachers can be supportively nonevaluative and the self can be harshly evaluative and critical.

In this diagram four horizontal bands represent four kinds of audience. Three vertical columns of dots represent three kinds of response those readers might give.

Table 13.1

The Spaciousness of Rhetoric in Terms of Audience and Response[2]

Audience	Sharing, but no response	Response, but no criticism or evaluation	Criticism or evaluation
Authority, that is, teachers, editors, supervisors, employers	●	●	●
Peers	●	●	●
Allies: readers who particularly care for the writer	●	●	●
Self only: private writing	●	●	●

I would like to think of this model as adding some useful enrichment to the useful models of writing that Jim Corder provided in his textbook (*Contemporary Writing* 326-29). But he was way ahead of me in calling attention to the distinction that I'm stressing here: the effects on the writing process that follow from whether or not the writing was "done to satisfy the requirements and expectations of particular occasions and assignments" (322).

If we pay more attention to the everyday, self-sponsored, independent writing that ordinary people do, we will build a more commodious and complex but more accurate model of writing. Most people—and even many scholars in composition—have taken for granted what I've called the dominant model because of the prominent and powerful experience of writing in school and in the workplace. Indeed—to be even more overt about my own agenda here—I suspect more people will do more self-sponsored writing when they are not seduced by the dominant assumption that people never write by choice unless they are "real writers" and that writing is always a painful struggle. A better model of writing should even improve the experience of teachers and students and people in the workplace as they negotiate writing under compulsion. That is, teachers will teach more effectively and students and people on the job will learn more effectively when they can free their minds from the dominant model and realize some crucial truths: It's common for ordinary people to write by choice and with pleasure, to keep some of their own writing private, to share some of it for the pleasure of sharing alone—getting no response, and to share some of it for nonevaluative response. (For some pedagogical implications, see my "Map of Writing.")

NOTES

[1]Perhaps some readers will want to exclude this last item from the category of everyday independent writing because it involves writing done for teachers. But it is relevant to my argument that a great many people are spending a great deal of their time and money in MFA programs to write things they don't have to write. MFA programs are one of the fastest-growing segments in higher education. Of course many of the participants hope to publish and become famous published "real writers," yet a fair number of them are happy to enroll in order to advance their agenda as self-sponsored writers. And many of those who start off hoping to become famous published writers soon or eventually recognize that they probably won't succeed—and yet keep on writing. A comparable but newer growth field in higher education is the undergraduate major or minor in writing—and elective writing courses. More and more high schools also have elective writing courses. Undergraduates in these majors may be more prone to the belief that of course they will end up famous published writers, but this certainly isn't

true for everyone in the proliferation of elective writing courses on so many campuses.

²My diagram would have been hard to read if I'd tried to map certain other important categories of audience:

- Are we writing to readers we know or to readers we don't know?

- Are we writing to a large group or a small group or just one reader?

- Are we writing to absent readers or to readers who are present with us as we write (as is so common in writing classes and not infrequently on the job)?

WORKS CITED

Applebee, Arthur N. *Writing in the Secondary School: English and the Content Areas*. Research Report No. 21. Urbana, IL: NCTE, 1981.

Barton, D. *Literacy: An Introduction to the Ecology of Written Language*. Cambridge, MA: Blackwell, 1994.

—. "The Social Nature of Writing." *Writing in the Community*. Ed. D. Barton and R. Ivanic. Newbury Park, CA: Sage, 1991. 1-13.

Barton, D., David Bloome, D. Sheridan, and Brian Street. *Ordinary People Writing: The Lancaster and Sussex Writing Research Projects*. Number 51 in the Working Paper Series, published by the Centre for Language in Social Life, Department of Linguistics and Modern English Language, Lancaster U, Lancaster England, 1993.

Barton, D., and R. Ivanic. *Writing in the Community*. Newbury Park, CA: Sage, 1991.

Barton, D., and S. Padmore. "Roles, Networks, and Values in Everyday Writing." *Writing in the Community*. Ed. D. Barton and Ivanic. Newbury Park, CA: Sage, 1991. 58-77.

Bloome, David. "Introduction: Making Writing Visible on the Outside." *Ordinary People Writing: The Lancaster and Sussex Writing Research Projects*. Ed. D. Barton, David Bloome, D. Sheridan, and Brian Street. No. 51 in the Working Paper Series, published by the Centre for Language in Social Life, Department of Linguistics and Modern English Language, Lancaster U, Lancaster England, 1993. 4-11.

Burrows, David. *Sound, Speech, and Music*. Amherst: U of Massachusetts P, 1990.

Buell, E. H., Jr. "Eccentrics or Gladiators? People Who Write about 'Politics in Letters-to-the-editor.'" *Social Science Quarterly 56* (1975): 440-49.

Chomsky, N. *Reflections on Language*. New York: Random, 1975.

Corder, Jim W. *Contemporary Writing: Process and Practice*. Glenview, IL: Scott, Foresman, 1979.

—. "Hunting for *Ethos* Where They Say It Can't Be Found." *Rhetoric Review* 7.2 (Spring 1989): 299-316.

DiMatteo, Anthony. "Under Erasure: A Theory for Interactive Writing in Real Time." *Computers and Composition* 7 Special Issue (April 1990): 71-84. Online. Internet. http: //corax.cwrl.utexas.edu/cac/Archives/v7/7_spec_html/7_spec_5_DiMatteo.html

Duin, A. H., and C. J. Hansen. *Nonacademic Writing: Social Theory and Technology*. Mahwah, NJ: Erlbaum, 1996.

Elbow, Peter. "In Defense of Private Writing." Written Communication 16 (1999): 139-79. Rpt. in *Everyone Can Write: Essays Toward a Hopeful Theory of Writing and Teaching Writing*. New York: Oxford UP, 2000. 257-80.

—. "Taking Time Out from Grading and Evaluating While Working in a Conventional System." *Assessing Writing* 4 (1997): 5-27. Rpt. in *Everyone Can Write: Essays Toward a Hopeful Theory of Writing and Teaching Writing* as part of the essay titled "Getting Along Without Grades—and Getting Along With Them Too." New York: Oxford UP, 2000. 399-421.

—. "The War Between Reading and Writing—and How to End It." *Rhetoric Review* 12.1 (Fall 1993): 5-24.

Elbow, Peter, and Pat Belanoff. *Sharing and Responding*. 3rd ed. New York: McGraw, 1999.

Faigley, Lester, and T. P. Miller. "What We Learn from Writing on the Job." *College English* 44 (1982): 557-69.

Feldman, C. F. "Two Functions of Language." *Harvard Education Review* 47 (1977): 282-93.

Finders, M. J. *Just Girls: Hidden Literacies and Life in Junior High*. New York: Teachers' College P, 1997.

Gere, Anne Ruggles. *Writing Groups: History, Theory, and Implications*. Carbondale: Southern Illinois UP, 1987.

Graff, H. *The Literacy Myth: Literacy and Social Structure in the 19th Century City*. New York: Academic P, 1979.

Graves, Donald. *Writing: Teachers and Children at Work*. Portsmouth, NH: Heinemann, 1983.

Hamilton, M., D. Barton, and R. Ivanic. *Worlds of Literacy*. Toronto: Ontario Institute for Studies in Education, 1994.

Howard, U. "Self, Education, and Writing in Nineteenth-century English Communities." *Writing in the Community*. Ed. D. Barton and R. Ivanic. Newbury Park, CA: Sage, 1991. 78-108.

Noddings, Nel. *Caring*. Berkeley: U of California P, 1984.

Odell, Lee, and Dixie Goswami, eds. *Writing in Nonacademic Settings*. New York: Guilford, 1985.

Schweickart, P. P. "Reading, Teaching, and the Ethic of Care." *Gender in the Classroom: Power and Pedagogy*. Ed. S. L. Gabriel and I. Smithson. Chicago: U of Illinois P, 1990. 78-95.

Sheridan, D., Brian Street, and David Bloome. *Writing Ourselves: Mass-Observation and Literacy Practices.* Cresskill, NJ: Hampton P, 1999.

Stotsky, Sondra. "Participatory Writing: Literacy for Civic Purposes." *Nonacademic Writing: Social Theory and Technology.* Ed. A. H. Duin and C. J. Hansen. Mahwah, NJ: Erlbaum, 1996. 227-56.

Tebeaux, E. "Nonacademic Writing into the 21st Century: Achieving and Sustaining Relevance in Research and Curricula." *Nonacademic Writing: Social Theory and Technology.* Ed. A. H. Duin and C. J. Hansen. Mahwah, NJ: Erlbaum, 1996. 35-56.

14
Weaving a Way Home: Composing a Personal Geography

John Warnock
University of Arizona

At any rate, I proposed what might or might not be a geography course.

We are always situated somewhere when we speak, whether or not we have placed ourselves deliberately.
—Jim W. Corder (*Yonder*)

In Jim Corder's last book (*Yonder: Life on the Far Side of Change*—a book he wanted to be "a scholarly sort of book written in a personal sort of way" (x)—one of the pieces in that book is titled, engagingly, "I Proposed a New Geography Course, but the Curriculum Committee Turned It Down." In the essay Corder imagines a great shambling course that would

> start with some questions and puzzles and issues: How and why do we acquire so much misinformation about places and their relationships? How do we rhetorically construct a geography for ourselves . . . ? Why do some people become attached to places while others don't? When and why do we humans begin to think about place? What are our personal geographies? (129)

As interesting as we may find Corder's questions, I imagine we can understand readily enough why the committee might have turned down his proposal for what he called this quaintly interdisciplinary course. Those in literary studies would have wondered what the texts for the course were and perhaps been concerned about the excessively local focus of an inquiry into "place." Those in area studies, though they would be untroubled by any allegations of regionalism, might have found his emphasis on place excessively local as well and like their colleagues in literature would have found his emphasis on the personal troubling. Creative writers would have been troubled by the apparent absence of issues of craft and perhaps by the apparently nonfictional emphasis of the inquiry. The geographers (should there have been any on the committee) would have found the idea of a personal geography oxymoronic.

And what of the rhetoricians on the committee? Might they have spoken up in support? If they had done so, how might they have needed to couch their arguments? This chapter is a roundabout suggestion of a way they might have done this. It's also a suggestion of a way that those of us who wish to say we are rhetoricians might think about what a rhetorical approach to a subject entails.

The same year that *Yonder* was published, I proposed to the director of the Bread Loaf School of English, Jim Maddox, that I teach a course at the Bread Loaf campus in Santa Fe, New Mexico, that would be called Cultures of the American Southwest. I had been teaching at Bread Loaf for several years, at the Vermont campus the first year, then at the New Mexico campus, most often a course called Travel Writing, and one year an Introduction to Rhetorical Theory.

The course I was now proposing, though it was to be called a course in cultures, would be offered, as my other courses had been, as part of Bread Loaf's upstart School of Writing rather than its established School of Literature.[1]

I wasn't entirely sure what this course would turn out to be, or what it should be called, any more than Jim Corder seems to have known this about his course. But I think Jim would have approved of my course as it has turned out. I think he would have found that it has raised some of the questions he wanted to raise and in the way he would have wanted to raise them. It could well be called a kind of personal-geography course. It could also, I want to argue here, be called something Jim might have not anticipated, even though it would have reflected a lifelong preoccupation of his. It could be called a rhetoric course.

It is a commonplace (among academic rhetoricians, at least) that rhetoric is situational. That claim is usually understood to begin with Aristotle, who defined rhetoric as "the faculty of discovering the possible means of persuasion in reference to any subject whatever." At an early moment in the recent recovery of rhetoric in the academy, the claim was asserted as fundamental to rhetoric by Lloyd Bitzer in "The Rhetorical Situation." Bitzer's claim has been amplified and revised but not fundamentally challenged by later commentators. Probably a great many of us these days answer the question, "So, what's rhetoric then?" by saying that rhetoric is an approach to language use that calls for paying special attention to situation.

The claim that rhetoric is situational is one reason why the practice of rhetoric is only uneasily at home in the academy. So conceived, this rhetoric challenges a fortiori the form/content assumption that remains at the center of academic practice—the assumption that knowledge is the content the academy exists to generate (or produce, or discover) and that this knowledge is contained within a particular form only so that it may be disseminated (communicated, transmitted, delivered, passed along). Essential to this idea of knowledge is that it is precisely that which transcends situation, if we mean by situation something contingent, limited, local, personal. In the language of the Platonic scheme, we could say that the academy is understood to sponsor *episteme*, rather than the merely situational *doxa*.

Some practitioners of cultural studies have developed a concept of situated knowledge as a feature of their critique of academic hegemony. So far, they have remained chary about throwing in explicitly with rhetoric, however, as have others whose arguments seem to be dragging them in that scandalous direction, as would anyone who wished to make his or her way in the academy today. We see this chariness, for example, in Donna Haraway's *Modest_Witness@Second_Millenium.Female Man©_Meets_Oncomouse™: Feminism and Technoscience*. The book challenges the assumptions of normal science, and by implication normal academic discourse, and opens up a space for rhetoric. But the word *rhetoric* occurs only occasionally and is always used to characterize some limited argumentative move rather than as a candidate for a characterization of what discourse in this newly opened space must necessarily be. In Terry Eagleton's *Literary Theory*, there is in the last chapter a famous instance of rhetoric being claimed explicitly to be a more appropriate name for a form of normal academic discourse that typically eschews the term—namely literary criticism. But it is one thing to claim that others practice rhetoric. It is another to discourse as if one believed it to be true of oneself.

The emphasis on situation might be, then, not just a new idea but an idea that scandalously undermines one's claims to academic legitimacy. And yet, at the same time that rhetoric and composition (and cultural studies) have been asserting the importance of situation, they have also been consolidating their position in the academy. One of the ways we in rhetoric and cultural studies have made the scandal of situatedness acceptable within the academic paradigm is to have treated situation itself as a given, as a kind of content that amounts to our *physis*, our nature—to be analyzed, not interpreted, or even worse, constructed. With situation as our foundation, we may decorously dispute within the playground of the academy what might or might not be appropriate to the situation.

We tend to do this even when the discussion purports to accept the claims of what has been called epistemic rhetoric, the claim that (all?) knowledge is constructed. How do we get away with that? We get away with it by asserting such claims in the *style* of normal academic discourse, which, however varied it might be, is in any case precisely a style that is suited to representing knowledge as what the academy takes it to be, that is, as nonsituated. We tend to call for contextualization in language that is not itself contextualized. In the last few decades, proponents of the two disciplines that have been most busy asserting the situatedness of knowledge and discourse—cultural studies and rhetoric—

have found a comfortable home in an academy whose practices remain founded on a mission understood to be the production and dissemination of (universal) knowledge. They did this by producing discourse the quality-control workers of the institution recognized as academic. From the fact that arguments about situatedness have not interfered with the academic success of those making them, we could reasonably conclude that the style is the substance in this affair, that as long as apparently radical challenges to the academy are made in the normalized style of academic discourse, guardians of the institution will not need to worry about maintaining the status quo. This argument might discourage those who wish for radical change. It might encourage those who find in the academy something worth preserving. It could leave us feeling impatient, on the one hand, with the timidity of academic radicals and, on the other, with the self-righteousness of those academic radicals who are careful to produce their challenges to the status quo in ways that preserve and enhance their status in academy, all the while condescending to anyone who purports to value what the academy does.

So discourses that assert the essential importance of situation can be and often are belied by their styles. We might then ask what does an actually situated discourse—one that might not play so well in the academy—look like?

By implication, we might look for such discourse in the discourses that are defended against in the academy by being found wanting in particular ways. Here I'd like to consider three of these genres, or aspects of discourse, all of which are marked as *infradig* or otherwise iffy in the academy: the personal, the regional, and the written. As it happens, all three are precisely what I hope my course will sponsor.

"The personal," like "the rhetorical," is something that is often to be found marked in the realms of official discourse as "the merely." I recently heard a colleague present at a forum in the English Department that has been instituted by our Head, Larry Evers, to encourage exchange across the specialties.[2] She is a very accomplished literature scholar and feminist who has published several books of criticism. She read the introduction to something she is working on now that begins with an account of a fourteen-year-old girl (herself, described in the third person) finding some journals written by an ancestor from the American South who had owned slaves and lived through the Civil War. The account was staged as a way into an inquiry about historical and cultural realties. It was not a memoir. Yet in the discussion, the first question, from a just-hired assistant professor, had to do with how, given the fact that she was working with family documents, she was to achieve critical distance. In her response she said that that was of course a crucial question, that she had read us the most personal part, and that she would be taking care not to engage in the indulgence of personal writing.

In all of this, although she had taken a risk in writing from family history, and consciously taken it, she was in the end reproducing the system of values that underlies normal academic discourse in English or at least in literature, defending against the personal in a way that is standard in this discourse. Interestingly, a recently hired writer of creative nonfiction (hired a few years ago) then raised his hand and said that the personal was exactly what he had found most compelling, and he wanted more of it. In any case, the exchange nicely demonstrated the dubious position of the personal in normal academic discourse.

When it comes to the province of the local or the regional, it is obvious how low on the pole of prestige such concerns are in an academic context. The local and the regional escape being excluded from the domain of normal academic discourse by being exemplified or museumized, and thus capable of being theorized. When the local or regional is taken to offer an example of some larger phenomenon or when it is seen as an artifact of something larger than itself, it becomes academically decent. Folklore, for example, is fine to the extent it is thematized or universalized. To the extent that it is only lore of a particular folk, it is academically suspect.[3]

A similar movement takes place in the assertion that the personal is the political. The personal has no cachet in academic discussion in its own terms. If it can be translated into the political, it can enter the academic conversation. Notice that within the academy, there is less of an impulse to make the apparently symmetrical claim that the political is the personal. Movement in that direction makes the political count for less, not more, in normal academic discourse.

At one point in the history of modern English studies, American and other Area Studies had to struggle against this attribution of less-than-academic status. Both are now an unremarkable presence in the academy and respectable in its terms. To achieve this status, they had to show that they could conduct their business in the style of the academy, just as rhetoric and composition and cultural studies did later. Defenders of normal academic discourse would say that this showed that these subjects were intellectually worthwhile. Those who claim to wish to subvert the academy might describe what they have done as "assuming the position." And yet even those who claim to be interested in subversion would find themselves unhappy if the academy began to define itself as concerned primarily with promoting the national (or some even more local) interest. Normal academic discourse doesn't do that, not consciously.

We should not allow ourselves to be misled by the powerful emphasis in universities like my own land-grant university on providing benefits to the local economy, culture, and so forth. If you can do that with soft money, you'll be allowed to do so as long as you want. But it won't get you tenure. Tenure comes not with a showing of local benefit but with a showing of a contribution to the store of (unsituated) knowledge.

Finally, the written. We risk scandal in the academy if we contend that knowledge is written in any sense other than "what is given"—the theological "It is written" sense. To see knowledge as written in that other sense, the sense of its being the product of composition, not merely of discovery, is beyond the pale. Analysis is the preferred activity and interpretation is to be avoided, if possible. In normal academic discourse, writing is said to be important, but only as a means of communication, that is, delivery of content. This is true even in the movement known as writing across the curriculum, where writing is promoted as learning, that is, as a means of discovering content. The idea that writing is constitutive is usually beyond the pale.

In the last year, in the course of performing certain duties as a quality-control officer in my university, I served on three committees that had to read and evaluate the work of creative writers (one sabbatical award committee, and two posttenure review committees). As it happened, no representatives from the creative writing program were serving on any of these committees. All were humanists. It was interesting to watch the members of these groups address the material that had been submitted to them. They did not reject it—the program in creative writing at my university has been highly ranked for many years now, which no doubt helps here—but they sort of waved their hands around. Particularly in the sabbatical award committee, they admitted they didn't know how to evaluate the work proposed to be done. (In posttenure review, one can count creative publications just as easily as one can count scholarly ones, and let that stand for content.)

But it is not just creative writing that doesn't play easily in the academy. Composition doesn't either, or rather might not. Composition can be reduced to a concern with matters of style, with style understood as form, not content. Those who make style a matter of concern primarily for clarity are engaging in this reduction. This is not a move open to creative writers, whose style is always seen as implicated with meaning. But it is a move that tends to be well received in the academy, both by faculty and a fortiori by students. English professors may want to add to the equation a concern with *grace*. One must investigate further to see if this amounts to a serious recognition of the limitations of the normal academic view or simply a sense that style may act as an adornment for content.[4]

Similarly reductive is the view of rhetoric that sees it as the manipulation of language for the purpose of manipulating an audience to achieve preconceived ends. Under these assumptions, the aim of the discourse may be seen to be obfuscation and deception, rather than clarity (though speakers' motives are presumably clear to themselves at least). Rhetoric is something one may be accused of. For the purposes of this discussion, the important point is that in this view style is seen to be just as detached from the person of the speaker as it is in normal academic discourse. In this view rhetoric is simply the evil twin of normal academic discourse, not something that in its fundamental premises challenges normal academic discourse.

Writing seen as personal, as regional, and as written, then, is writing that we could call situated. It is situated not simply in being contingently local, or in accepting the rules of some particular game, but in having a significant relation to all three of the elements we've discussed here: to the writer, to the geographic place, and to the language.

What more can we say about the kind of relation that is imagined here? A few more observations about what kind of relation it isn't before I turn to what kind of relation I think it is.

Today it is common to refer to culture as if it is just sitting there, something that can be unproblematically contained, defined, packaged. You have this culture here and that one there and you can take your pick of which one to visit and perhaps study, and when the tour is over go home with your souvenirs, more cultured than before perhaps, more aware of diversity, perhaps, but unchanged in any way that matters much. We could call this relationship touristic. It is very common, and a good deal of writing approaches culture in this way, writing of the sort found in the airline magazines and the Travel section of the *New York Times*. This approach is common in classrooms, too, though in universities it tends to be dressed up in discourses that obscure the fact. In any case, courses that approach the teaching of culture as if it is a prepackaged object flirt with a kind of tourism, it seems to me. A writing course that focused on writing for the market could take this approach, would probably need to take this approach, in fact, given the nature of the market. I knew this wasn't the approach I wanted to take at Bread Loaf.

It is possible to approach an encounter with another culture in quite another spirit—in the hope that the other culture might supply something important that we find missing at home, and become home. We can be surprised by this hope. We come over Glorieta Pass on Interstate 25 and see the Rio Grande valley in the rain shadow between us and the Jemez Mountains, and we feel something stir deep in our chests and wonder if we might not want to move to this place, become one of them, go native, hoping to find or create the home that is our true home. We may say, as the writer Jim Sagel told me he did on first seeing northern New Mexico, "This is where I should have been born." The American Southwest, and Santa Fe in particular, is populated by a good many people to whom this has happened. We could call this the sentimental approach to culture. Developers and marketers make heavy use of it, but we should be careful not simply to sneer at it. It can represent an effort to open the door to revision of self, of world, a determination to make things better. Still, I was not interested in turning this course into a recruiting device.

But it's not as if one can avoid either the sentimental or the touristic relation to the situation (in this case, the American Southwest, and oneself in it) simply by developing somehow an account of the culture that represents the culture as it is. Proponents of cultural studies are very good at showing how particular representations fail to give us things-as-they-are and instead do various kinds of cultural work that may or may not, and usually do not, comport with what the representations claim to be doing. But in my opinion, proponents of cultural studies too often talk as if their own representations come, or might come, a whole lot closer to things-as-they-are, as if their own representations are not situated in the way they are so good at showing other representations to be. In any case, they leave unquestioned the notion that representing things-as-they-are is what we ought to be about, which is, I've claimed here, the fundamental assumption of normal academic discourse.

It is sometimes said, for example, in what I'm calling cultural studies, that since everything is always already ideological, one has an obligation to declare one's ideology in any discussion, that this keeps one honest. But is it imagined that such a declaration is any less subject to manipulation than keeping to oneself about such matters? Even more important, even assuming that such a declaration is sincere, is it not not a declaration of things-as-they-are but rather a utopian declaration of a wish-as-to-the-way-things-might-be just as much as any sentimental, idealistic account? "Rhetoric is never innocent," wrote James Berlin, a statement frequently quoted in arguments for the importance of cultural studies to rhetoric and composition. This would be important news only to those who were hoping rhetoric might "take its legitimate place in the academy," to do which it would have to be made innocent in just the way that normal academic discourse claims itself to be. Do these scandalizers of us rhetoricians assume that the discourse of cultural studies *is* innocent? They are not so rash as to say so explicitly. But the style of their analytic performances—and of their supposedly redeeming declarations of their own ideology—makes it reasonable to assume that they do.

Situated discourse of the kind I'm talking about here may address questions of what are the facts of the matter. But it also raises questions about what kinds of commitments representations of a subject—one's own representations as well as those of others—can be seen to make, deliberately or not, what ideas and ideals they stand for, deliberately or not, and what relations the representations have to those who make the representations, who are understood to be partly constituted by their own acts of representation.

As James Boyd White has pointed out in "Rhythms of Hope and Disappointment," "We the people" in the Declaration of Independence might more accurately have been written "We the white, land-holding, unenslaved males." But, he goes on, that account would also have been inaccurate—in failing to represent the aspirations those men had, or in any case were willing to represent *as* theirs, in a language that today puts us into a position to continue to struggle toward a state of affairs in which "the people" are not defined as narrowly as they were in that day. "We the people" is a vehicle for hope and action of a sort that the more accurate statement would not have supported.

The American Southwest I take to be a similar kind of construction, though no doubt a better word is *composition*. In composing it, accuracy (and its cognates: analysis, demystification, etc.) can be important, but it is not the only matter of moment.

For myself, the American Southwest is the region of my birth, but it has not until recently been a place I would have wanted to call home. For most of my life, it was a place I wanted to think of myself not as belonging to but as being from. It was not that I viewed myself as an interloper (as I and other Anglos might have been viewed in the eyes of the Hispanics who lived here, and as the Hispanics themselves might have been in the eyes of the Apaches, and as the Apaches in the eyes of the Tohono O'odham, and as all of the above might have been in the eyes of the Hohokam, had the people survived). I didn't know enough of the history of the region even to entertain that suspicion, nor was that the theme of such history as I was taught. This place belonged to me, all right, as I saw it: I just didn't belong to it.

I did know that the families of both my parents had come from the East. Their stories were the stories of many Anglo-Americans who now resided in the American Southwest. My father had been born in New York City, and his parents had come West when his father contracted tuberculosis; my mother had been born in Kentucky, and her parents had come West because her father saw it as a land of opportunity. Not surprisingly, I had a strong sense that the center or origin of things was not where I lived—there was no kiva in my village—but back there in the East, which of course *was* the theme of such history as I was being taught.

In high school I spent a summer as an American Field Service exchange student in Argentina, during which time I was often asked to speak for and explain America in ways I felt painfully unequipped and unentitled to do. But the experience made it even clearer what I needed to do to get equipped and entitled. So when I graduated from Tucson High School in 1959, I left the American Southwest to attend a classic liberal arts college in New England. As a public school kid from the West, this was clearly not a place I belonged either, though I could believe— most of the time—that it was a good place for me to be at the time. But in the origin story I was then inhabiting, there was a point of origin further East still, I knew, so on premises very similar to those that got me to Amherst, I went on to graduate school at Oxford in English Literature (nothing written after the 1900s, of course, except by special dispensation). For me, with Anglo, German, and Scotch-Irish grandparents, that was it: There was no place further East that one might go in search of what was authentic. So I began to swing back, first to law school in New York City, then to work as a law clerk in San Francisco, and, finally, when I decided not to practice law, to Laramie, Wyoming, to visit a friend, and stay a while.

I spent over twenty years in Laramie, most of that time at the University of Wyoming, which hired me after a professor got sick and I took over his classes, some of which were composition classes, and I got interested in the question of what composition was and might be. Not long after I arrived in Laramie, I began to see and speak of myself, as I had not before, as a Westerner. In this discussion what is important about this characterization is that it was less a matter of fact than it was something I was then choosing, or trying to choose, as true. I see now as I didn't then that I was looking for a place I could call—and know as— home.

We tend to think of home as a place, but it is not just geography. It is, in Corder's phrase, a personal geography, a geography to which we have a kind of committed relation. It *is* a geography and not just a place, however, and the etymology of the word *geography* tells us that we are dealing not with actual places but with representations of them, written worlds. In this way we might be reminded that whatever else home is, it is inescapably a symbolic construction.

We tend first of all to think of a home as a *physical* place, but we also can think of it as a conceptual one, as when we speak of a disciplinary home, in which case we may focus more on home as a place for play, in the sense of the term adumbrated by Johann Huizinga in *Homo Ludens*. At Wyoming I hoped to find a physical home, but in the years I was there, I also was engaged in trying to find, or make, a disciplinary home. After Oxford I went to law school because I knew that I would not be able to be at home in English as it was then instantiated in American (or European, for that matter) graduate programs. I didn't expect to find a home in the practice of law, however, and I didn't. But at Wyoming, in the early 1970s, teaching four composition classes as a part-time instructor and then as Director of Composition, I began to see the lineaments of what might be for me an academic home, a place where the primary concern was with writing (as distinct from interpretation of the written or from production of writing of the sort called "creative"). This home was not at that time an existing structure one might simply move into. It was something more like a hope, and if it became an actual home, it would have to be constructed out of the materials at hand (the work of composition) and other academic materials that might be brought to the project, such as the then largely ignored history of rhetoric in Western education, and developments in social sciences, particularly, at that time, linguistics and psychology. Since this vision of possibility was not shared by my colleagues in the English Department at Wyoming, it was necessary to search elsewhere for help with construction. Help showed up regularly at barn-raising parties like the Wyoming Conference on Freshman and Sophomore English,[5] the Rhetoric Institute,[6] and the Wyoming Writing Project for the teachers in Wyoming's schools.

Those of us who wish to claim a place in the academy must a fortiori identify an academic home. Even joint appointments elect a home department. We may or may not understand ourselves to have a personal relation to this home, though I've noticed that many defenses of it— against the inroads of interdisciplinarity, say—often have an animus that can get personal rather quickly.

I don't believe that everyone, in the academy or outside it, wants to have a particular physical place they call home, or even a disciplinary one, and I don't know if it is a good or bad thing that I have wanted and thought I might hope to have one. Is it not better to be, like Diogenes the Cynic, cosmopolitan, a citizen of the world? Perhaps it is, sometimes, for some people. Certainly I can remember the exhilaration of seeing myself in these terms as I left Tucson for college and other points East in 1959.

Furthermore, is not a feeling for home a form of ethnocentrism, and isn't that a bad thing? Certainly it can be, though in contemporary popular culture, we tend to consider the profound attachment to particular places among, say, the Navajo and the Pueblo people as a kind of cultural virtue. There is good reason, moreover, to think that we have no choice but to be ethnocentric, and thus that it may be more responsible to take our own ethnocentrism as our point of departure.[7]

A desire for a home is not without political consequences. It's obvious that there are powerful forces working in the contemporary global economy to detach us from the places we might want to call home, unless perhaps we are content to see ourselves as being at home in the free play of the signifier or the market. In such a situation, the hope of home can be a serious inconvenience for the one who has the hope. I can remember when it dawned on me that any reluctance to move had a price in the academy and that the most realistic view of the value of one's scholarship was that it allowed one to move. A corollary is that from the point of view of those who conduct the affairs of the global economy, claims of home are often subversive. The people of Chiapas have shown this in their resistance to the consequences of the North American Free Trade agreement.

In the end, Wyoming was not a place I could call home (you can't always get what you want), and in 1990 Tilly Warnock and I left Wyoming and came to the University of Arizona in Tucson. We came here not because we were trying to get to Tucson, or return to a homeland (if we'd wanted to do that, we might have ended up in Columbus, Georgia, where Tilly was born) but because the University wanted Tilly to direct its Composition Program, and she wanted to do it, and they had a place for me too. But in coming here I've found a place that I can now call and know as home.

I don't have a clear sense of the extent to which this has happened because I was born and spent my first eighteen years here. That no doubt counts for something, of course, though when we came here, I had spent almost twice as many years elsewhere as I had here. In any case, I have found myself responding in deep and resonant ways to the landscape here and to the history, the people, the cultures, the languages, and the local institutions. Since we arrived in town, I've been developing a personal geography that has not only a spatial and historical dimension but one that places me in a kind of committed conversation with everything, including the land and the university. Teaching the course at Bread Loaf has allowed me to develop that multidimensional conversation in manifold ways and has been a great blessing for me.

As for my students, I of course cannot assume that the American Southwest will turn out to have anything like the status of home for them. The region has a kind of glamour in contemporary popular culture, but it could turn out to be a geography that someone might find uncongenial (I need black dirt, dirt you can run your fingers through!), or infradig (If I see another howling coyote. . . !), even appalling (Oven mittens for the car?). But what I do want to insist on is that whatever the students turn out to make of their encounters with the region and various representations of it be something to which they have a personal relation, for which they feel personally responsible, not only responsible in the ways allowed by normal academic discourse. This personal responsibility entails developing an idea of it that recognizes that many different kinds of stories can be told about it, including the stories of its past as the dwelling place of those sometimes called Native Americans, the stories of its past as part of New Spain, and the stories of it as a site of military-industrial adventure (as I like to call it). It entails understanding something of what is at stake in the stories being told in this way or in that, including those stories told in the form of normal academic discourse. I don't want them to study the American Southwest only as an objective phenomenon, as something entirely other, something they might represent without in some way being personally implicated in this representation.

This is not quite the same as asking the students to be self-reflexive, the gesture that seems to be thought to have some power to inoculate the writer or the reader against bias or ideology or ethnocentrism, or somehow to exorcise these things, again making the world safe for normal academic discourse. Still less is it asking them to say "what the American Southwest means to me" or to recount "the adventures of their soul" in, say, Monument Valley. That stance toward the project discounts far too deeply the fact that transactions like these are constructed from and enabled by language, and thus may be constructed otherwise than they have been.

A way of summing up these attitudes might be to say that I'm asking that we consider the American Southwest not as a region or a culture but as a rhetorical situation. Among the classical rhetorical concepts that have been drawn upon by contemporary commentators on the rhetorical situation, two that might help students locate themselves in relation to what I'm asking them to do as readers and writers are the ideas of *stasis* and *kairos*. *Stasis* is the undertaking to discover what assumptions are shared with one's interlocutors, so that one can understand also what is at issue. In our own time, Kenneth Burke reminded us of the necessity of locating this common ground in his claim that identification is necessary to conflict.[8] *Kairos*, a term that was recovered for contemporary rhetoricians by James Kinneavy, is defined by him provisionally at the beginning of his essay on the term as "the right or opportune time to do something, or the right measure in doing something." A focus on *stasis* and on *kairos* might help students get a sense of what it could be to situate their discourse and to begin to develop their representations of the American Southwest not as something simply given, nor, on the other hand, as something simply under erasure, but as something under composition, something they will choose.

Notice that neither of these ideas—*stasis* or *kairos*—has an acknowledged place in the practice of normal academic discourse, and neither has an acknowledged place in the discourses of cultural studies that adopt the style of normal academic discourse. When we purport to be dealing in the realms of knowledge, as distinct from opinion, of content as distinct from form, we have no need of inquiring into matters of *stasis* or *kairos*. It's not that such questions cannot be asked of normal academic discourse: It's that once we are professionalized into the academy, these matters tend to become unavailable for analysis. In comprehensive examinations and dissertation defenses, Tilly sometimes asks candidates why they think a particular set of ideas—ideas about hegemony, or the circulation of cultural capital, let's say—have become so salient in certain contemporary academic discourses. It is interesting to see how often candidates, even candidates in our graduate program in rhetoric and composition, seem to assume that the only sensible answer to such a question is that these ideas are true.

I ask students to write, then, as if it matters how one represents the American Southwest (or anything else) and that we have a choice in how we do this. Just how it matters and to whom is part of what may need to be discovered. But it should be seen to matter in any case to the one making the choices. The geography here is necessarily personal geography. Finally, however, I would prefer to call it "rhetorical geography," on the understanding that the rhetoric of the representation has consequences not just for the audience—that is, the choices are not simply strategic choices—but also for the person who chooses to make this representation instead of that one. I think Jim Corder would accept this revision, but I'm not sure his committee would have seen it as much of an improvement.[9]

NOTES

[1]The School of Writing, which Dixie Goswami helped to install as part of the School of English, is not to be confused, though it often is, with Bread Loaf's glamorous Writer's Conference, which concerns itself exclusively with the kind of writing often called "creative," and which operates independently of the School of English.

[2]At the University of Arizona, the four official specialties are recognized in the graduate programs in Literature; Creative Writing; English Language and Linguistics; and Rhetoric, Composition, and the Teaching of English.

[3]In this way literary study leans against the wind in ways that are not sufficiently recognized. Many professors of literature in the American academy still take the position that particular works are primary, and resist any processing of these works according to one or another theory. It's possible to see this stance as supporting the local as against the cosmopolitan, and therefore as itself unacademic. Literature professors have fashioned a kind of escape for themselves by claiming to represent Literary Value or Western Literature. It's a tenuous out: Traditional literary study can be much more of a radical critical activity in the academy than it is sometimes given credit for being.

⁴A provocative treatment from another angle of a number of the issues I'm trying to deal with here is found in John Trimbur's "Essayist Literacy and the Rhetoric of Deproduction."

⁵This conference was started in 1972 by Professor Art Simpson, the then Director of Composition. In the 1970s and 1980s, especially under the directorship of Tilly Warnock, it was one of the premier gathering places for the workers and dreamers who now can be seen to have been constructing the academic geography of rhetoric and composition. It continues today, with a somewhat different agenda, as the Wyoming Conference on English.

⁶Ross Winterowd and I orchestrated this gathering in 1976 at the UW Science Camp in the Snowy Range in an effort to bring together all the people we could think of who seemed to be interested in making a place for (what might come to be called) rhetoric and composition in the Academy. Among those who were able to attend, besides ourselves, were Frank D'Angelo, Janice Lauer, E. D. Hirsch, George Yoos, and Richard Young.

⁷For a provocative defense of this position, which is of course also exceedingly inconvenient for normal academic discourse, see Richard Rorty, "On Ethnocentrism: A Reply to Clifford Geertz."

⁸See, for example, Burke's *A Rhetoric of Motives,* 19-28.

⁹The most recent syllabus for the course I have taught at Bread Loaf in Cultures of the American Southwest is available at http://www/ic.arizona.edu/johnw/sw.

WORKS CITED

Aristotle. *The "Art" of Rhetoric.* Ed. and Trans. John Henry Freese. Loeb Classical Library: Cambridge, MA: Harvard UP, 1926.

Bitzer, Lloyd. "The Rhetorical Situation." *Philosophy and Rhetoric* 1 (Jan. 1968): 1-14.

Burke, Kenneth. *A Rhetoric of Motives.* Berkeley: U of California P, 1950, 1969.

Corder, Jim W. *Yonder: Life on the Far Side of Change.* Athens: U of Georgia P, 1992.

Eagleton, Terry. *Literary Theory: An Introduction.* Oxford: Blackwell, 1983.

Haraway, Donna. *Modest_Witness@Second_Millenium.FemaleMan©_ Meets_Oncomouse™: Feminism and Technoscience.* New York: Routledge, 1997.

Huizinga, Johann. *Homo Ludens.* London: Routledge & Keegan Paul, 1949.

Kinneavy, James. "*Kairos*: A Neglected Concept in Classical Rhetoric." *Rhetoric and Praxis: The Contribution of Classical Rhetoric to Practical Reasoning.* Ed. Jean Dietz Moss. Washington, DC: Catholic U of America P, 1986. 70-105.

Rorty, Richard. "On Ethnocentrism: A Reply to Clifford Geertz." *Objectivity, Relativism, and Truth.* Cambridge: Cambridge UP, 1991. 203-10.

Trimbur, John. "Essayist Literacy and the Rhetoric of Deproduction." *Rhetoric Review* 9 (1990): 72-86.

White, James Boyd. "Rhythms of Hope and Disappointment." *From Expectation to Experience: Essays on Law and Legal Education.* Ann Arbor: U of Michigan P, 1999. 3-4.

15

Who Owns Creative Nonfiction?

Douglas Hesse
Illinois State University

∂My title asks an impolite and calculatedly agonistic question. At one level, it concerns academic turf, a quiet contest between creative writing and composition programs over the right to teach certain courses, claim certain genres, hire certain faculty, possess a certain curriculum. But at another level, the question proves useful for raising broad issues in rhetoric and composition studies.

Creative nonfiction serves as an umbrella term for a host of genres, including personal essays, memoirs, autobiographies, new journalism, and certain traditions of travel writing, environmental writing, profiles, and so on. Its current famous practitioners include writers like Annie Dillard, Joan Didion, Scott Russell Sanders, and John McPhee. Lee Gutkind, founder of the magazine *Creative Nonfiction*, claims too presumptively to have coined the name. Other terms have been proposed: *literary nonfiction* (adopted by Chris Anderson), or "personal journalism, literary journalism, dramatic nonfiction. . . the literature of fact" (Cheney 2). But *creative nonfiction* has stuck, especially where it seems now most to count: in MFA programs, in creative writing contests, and in the "how to be a writer" industry churned by Writer's Digest, Inc.

As everyone notes, the term is twice problematic, defining something by what it is not and separating some works as creative work from others that are not. I will note that many fiction-writing programs further complicate the creative half of the definitional morass by distinguishing between literary fiction (David Foster Wallace or Margaret Atwood) and genre fiction (Tom Clancy or Danielle Steele) as they warn off students who aspire to best-selling wealth. Fascinating as those definitional issues are, I leave them mostly suspended from this chapter. (For example, the question of how many liberties a writer can take with her true story and still be writing nonfiction is a richly vexed ethical and practical issue.) For some, such as the Illinois Arts Council grantmakers, creative nonfiction must have a strong narrative element; others would add such qualities as having "an apparent subject and a deeper subject," a tension between "the urgency of the event" and the "timelessness of its meaning," a sense of authorial reflection, and "serious attention to the craft of writing" and a "governing aesthetic sensibility" (Gerard 7-12).

The 1999 AWP (Associated Writing Programs) *Guide* lists thirty-five MFA programs in nonfiction. This number seems fairly unremarkable until one considers that in 1992 the *Guide* did not contain a single separate listing for creative nonfiction programs. It's obviously reductive to view MFA programs as representing the state of writing in the American academy. AWP has some three-hundred-member schools, and even among them are obviously more with writing and creative writing programs than have MFAs; beyond that are hundreds of departments teaching creative writing that are not AWP members.

Still the growth in less than a decade from zero MFA programs in creative nonfiction to thirty-five is metonymic of the broad growth of academic creative nonfiction. During this time period, university-based literary reviews (*The Iowa Review, The Georgia Review, The Hudson Review*), which had long sprinkled essays among poems and stories (often writers' comments on their craft or experiences, often discussions of genres or writers), increasingly devoted space to memoirs, personal essays, and experientially grounded nonfiction. Even more significant, they devoted more visibility to these genres, announcing theme issues and opening contests in creative nonfiction alongside earlier ones in poetry and fiction. Concurrently, the realm of summer writer's conferences, workshops, and colonies expanded to include nonfiction genres and writers. The upshot is that, by 2000, advertising in the *Writer's Chronicle* or in *Poets and Writers* is well populated with advertising for programs, submissions, grants, and meetings in creative nonfiction, many of them underscoring its financial aspects.

The growth of creative nonfiction comes against a backdrop of generally burgeoning creative writing. A 1970 directory of creative writing programs sponsored by the College English Association noted sixty-eight "well-developed programs in creative writing," with twenty-six masters, twelve MFA, and one PhD program (Iowa) among them (Sears 5). Growth since then has been steady and dramatic, as shown through the AWP Guides:

TABLE 15.1

Degree-Conferring Programs in Creative Writing 1975-1998

Year	BA	BFA	MA	MFA	PhD	DA
1998	318	12	143	83	29	1
1996	298	12	134	74	28	2
1994	287	10	139	64	29	3
1992	274	9	137	55	27	3
1984	155	10	99	31	20	5
1975	24	3	32	15	5	1

(Fenza, "Brief")

Certainly "noncreative writing" outside composition has also grown during that time period. In terms of degree programs, however, the bulk of that growth has been in technical and professional writing, at both undergraduate and graduate levels. Creative nonfiction is also taught outside creative writing programs, in courses partially or solely dedicated, unofficially or officially named as such, within composition programs or the English major, most often as part of that conceptual abyss advanced writing or advanced composition. More on that later. My point is that in terms of explicitly laying claim to these genres, creative writing visibly has been at the surveyor's office while composition studies has been tending other fields.

But why? One reason, of course, is that it's reasonable for creative writing to do so. The kinds of aesthetic and imaginative moves that earmark fiction are possible in nonfiction, too. Annie Dillard asserts that "the essay can do everything a poem can do, and everything a short story can do—everything but fake it" (xvii). Writers have always produced journalism and criticism alongside their poems and stories, so if a program aims to develop creative writers, it makes sense to expand their oeuvres. A second reason is more survivalist, territorialist, or colonialist, depending on the spin one takes. In her introduction to the *Best American Essays of 1988*, Dillard claims:

> Other literary genres are shrinking a bit. Poetry seems to have priced itself out of a job; sadly, it often handles few materials of significance and addresses a tiny audience. . . . The short story is to some extent going the way of poetry, willfully limiting its subject matter to such narrow surfaces that it cannot address the things that most engage our hearts and minds. So the narrative essay may become the genre of choice for writers devoted to significant literature. (xvi)

Whether Dillard's assessments accurately represent traditional genres, it is true that the numbers of creative writers in workshops these days swamp the opportunities available to make a living or tenure through publishing literature. Several years ago I speculated that critics looking for new scholarly opportunities were delighted to open the territory of creative nonfiction. The same might be said of creative writers looking to publish or programs looking to consolidate their presence in the academy.

There is a third option, namely, that people working in creative writing programs perceived something important in these genres that did not find a clear home elsewhere in the academy. Whenever rhet/comp types have explained the place of creative nonfiction in the writing curriculum, they have tended to focus on how writing in these genres is useful as a means rather than as an end. In this view creative nonfiction has been an orphan genre waiting for an institutional parent. But has this always been the case?

HISTORY LESSONS

Imagine that you are currently an undergraduate at the University of Illinois at Urbana-Champaign and that you're concentrating in creative writing. How will your major be designated on your transcript? "Rhetoric." To major in poetry or fiction at Illinois is to major in rhetoric, your plan of study including one advanced course in expository writing, four courses in narrative or poetry writing, Shakespeare, a distribution of literature and history courses, and a separate minor. Given the fierce opposition between creative writing and composition as James Berlin and others have characterized it, this arrangement would seem unlikely. Actually, it is more accurately labeled "quaint," though in the best sense of that word, for the association of creative writing with rhetoric extends back several decades.

The first class given the label *creative writing* was taught in 1925 (Myers). Prior to then—and for a couple decades after—creative writing was certainly taught in the academy, if not extensively, but taught under the rubrics of rhetoric and composition. Consider a 1905 textbook *Composition-Rhetoric* by Stratton Brooks and Marietta Hubbard. The first seven chapters mix what would now be considered composition and creative writing:

I. Expression of Ideas Arising from Experience
II. Expression of Ideas Furnished by Imagination
III. Expression of Ideas Acquired Through Language
IV. The Purpose of Expression
V. The Whole Composition
VI. Letter Writing
VII. Poetry (6)

Several things are striking about these contents. The first three chapters, for example, deal with autobiographical writings and direct observations (the material of creative nonfiction); with poetry, fiction, drama, and other "imaginative" genres; and with reading-based research. The chapter on "Purpose of Expression" declares that "there are two general classes of writing—that which informs and that which entertains" (133) and goes on to explain that each of these aims of writing can be divided into several forms: Bain's familiar modes of discourse. The chapters on letter-writing and poetry are striking in contrast. The former is highly practical, as Brooks and Hubbard underscore letter-writing's crucial role in "securing desirable positions" and "keeping up pleasant and helpful friendships," while the latter appeals to "our aesthetic sense; that is, to our love of the beautiful" (171, 195).

Rhetoric and poetic cohabited courses as well as textbooks. The text list for the 1907 "Outline for Rhetoric I" at Illinois included "Fulton's Composition and Rhetoric"; Le Baron Briggs's book of manners, "School, College, and Character;" and, most interesting, "A novel or a volume of short stories," and "Parkman's Oregon Trail, or Thoreau's Walden, or any other approved book containing description" (Brereton 471).

Three decades later, another textbook, *Creative Writing for College Students*, begins with types of formal exposition, the exposition of a process, the classified summary, definition of the abstract term or idea, and so on. Its second part divides writing the informal essay from writing the short story. Nonfiction and fiction are conceptually linked, both growing out of standard composition practices. The authors aspire to

> stimulate young undergraduates to think in terms of whole creations, not of unrelated fragments, such as relative pronouns, attributive adjectives, finite verbs, correlative constructions, and topic sentences. In general, then, the point of view of this book is that most writing is essentially creative in intent. (Babcock, Horn, and English vii)

The declaration supports Myers's contention that creative writing emerged not simply as an alternative way of teaching different genres of writing but rather of "returning composition to its original intent" and escaping grammar-bound rote and formulaic approaches (109).

But where does one stake "original intent"? Textbook writers like Babcock were responding to composition as it had developed in the late nineteenth century, a new disciplinary field distinguished by two developments: an accelerated turn from rhetoric as argument toward exposition and a burgeoning teaching load for composition instructors whose pedagogies were shaped at least partly by the need to cope with large numbers of students. One of the features of "new composition" after 1870 was an increase in assignments that had students use their own experience and observations as the basis for writing. Robert Connors offers at least four reasons. First, such writings reflected the romanticism manifested in familiar essays written during the early nineteenth century by Lamb, DeQuincey, Hazlitt, and so on. Students were told for the first time that "their responsibility is to be original in their writing, and usually this meant writing from observation" (313). Second, the shift from oral to written discourse entailed that "the subjects of rhetoric would become smaller scale, more private, more personal." Third, the admission of women to previously all-male schools created a void in rhetoric classes, whose traditionally agonistic modes, "natural" among men, would have been unseemly to practice in mixed classrooms. Connors claims that "Personal writing, for whatever reason, was not a part of rhetoric for 2,400 years, and its admission to rhetoric corresponds exactly to the admission of women to rhetoric courses" (64). Connors begs the question, of course, since not all would agree that personal writing is "part of rhetoric."

The fourth reason is even more complex. Invention had traditionally included deploying the commonplaces and allusions acquired through a classical education that late nineteenth-century students less frequently brought to the university. "In a world where wide reading—especially in the classics—could no longer be assumed," teachers sought appropriate writing tasks, and personal writings would suit (Connors 308). One hundred years later, of course, the sufficiency of classical invention would be assailed from a theoretical quarter rather than a practical one.

Whatever the combination of causes, writing was a largely undifferentiated terrain in the early part of the century, as composition formed itself following the demise of oratory. Even as an increasingly Fordist academy embraced exposition, belletristic writing continued to have a place within composition. Genres that would later be folded into creative nonfiction functioned as bridges between traditional composition and fledgling creative writing, often in the space of common courses. However, as creative writing programs emerged in the 1970s, they initially embraced fiction and poetry. This was at least partly for reasons of self-definition. Robert Scholes, James Berlin, and others have noted that writing is historically regarded as less important than literature in most English departments. At work is a double binary: poetic versus rhetoric and reading/consumption versus writing/ production. Scholes notes that historically texts labeled as *literary* merit attention whereas "non-literature is perceived as grounded in the realities of existence, where it is produced in response to personal or socioeconomic imperatives and therefore justifies itself functionally" (6). Drawing their boundaries at the traditionally literary frontiers of fiction and poetry, creative writing programs were able to avoid at least part of the binary by affiliating with poetics, although the student production of "pseudo literature" receives relatively little respect. Composition, even worse, is diminished as concerned with producing pseudo-nonliterature (Berlin 13).

The result was that in the 1970s and early 1980s, nonfiction was left mainly to composition studies and, to some lesser extent, to journalism. Creative nonfiction genres, especially the personal essay, were supported by Romantic theories of writing and by process pedagogies. They were mainstays of composition readers, especially the narration and description sections of those ubiquitous anthologies organized by modes. A popular book in the last decade explained "How the Irish Saved Civilization." It is not too extreme, in like kind, to claim that during the 1970s the *Norton Reader* and its many competitors saved the essay specifically and belletristic writing generally within the academy.

Despite the presence of imaginative readings and exercises, it's unreasonable to claim that creative nonfiction per se has thrived in composition courses. "Personal writing" has existed mainly as a mode versus as a genre, rather as a vehicle toward better writing than as a destination genre for students treated as serious writers. Under those process pedagogies grounded in a personal epistemic rhetoric (as distinguished from those process pedagogies grounded in current-traditional rhetoric), students were not asked to produce genre artifacts but mainly expected to experience how writers work. The exceptions were advanced writing courses aimed at English majors or writing minors, housed within composition programs, distinct from creative writing. The problem of advanced writing as a floating signifier has been well discussed. The rubric encompasses courses in business and technical writing, stylistics, writing for professional or disciplinary purposes, and so on. Among them were belletristic writing courses, perhaps as the only formal enclave of such discourses within English departments.

This was true, for example, at the University of Iowa in the 1970s, where 8W:100 and 8W:109, Expository and Advanced Expository Writing, tended to emphasize stylistic experimentation and essayistic writing, and where a course in The Art of the Essay helped identify the core of a Master's in Expository Writing developed in the later part of the decade. By the early 1990s, however, that degree had been changed to an MFA in Creative Nonfiction, a program curiously not affiliated with Iowa's celebrated Writer's Workshop.

Earlier I sketched the explosion of formal creative nonfiction programs as tracked by AWP in the 1990s. But this development was well underway in the eighties. In 1982-83 AWP inaugurated its award for the best volume of creative nonfiction, some nine years after beginning its award for poetry. By 1986 AWP minutes record that "the fastest growing creative writing programs are in nonfiction" (Rose). Was it the case that creative writing programs were poaching on rhetoric/composition's turf? Or was it the case that creative writing programs were expanding into genres heretofore unclaimed in the academy? Of course, to ask this question is to attribute strategic motives to national disciplinary entities, and it's a little silly to imagine AWP or CCCC war rooms complete with maps and stick pins.

Perhaps the best that can be said is that the status of creative nonfiction was simultaneously built in parallel spheres. Within creative writing programs, the issue was how to support writers working in these genres and others. With composition studies, curiously, the issue was more fundamental, even retrograde; it concerned the literariness of these works. For example, a landmark three-hour CCCC session in 1987 featured essayists Richard Selzer and Gretel Ehrlich, along with likely suspects Peter Elbow, Don McQuade, Robert DiYanni, and Chuck Schuster. Its focus: "Reclaiming the Essay as Literature" (*38th Annual* 77), a mission conducted in various other venues, including a conference at Seton Hall University which that same year produced Alexander Butrym's edited volume *Essays on the Essay*.

The late 1980's rise of the personal essay signals a powerful tension within composition studies. The social-constructivist movement was in full ascent with Pat Bizzell, Ken Bruffee, Lester Faigley, James Porter, and, of course, James Berlin all contesting ideas of authorship and the immanence of the aesthetic. And yet, at the very same time, seemingly in another disciplinary province, people were seeking to elevate nonfiction genres precisely by claiming their literariness. Jim Corder mused that "we might one day learn how to study great works of non-fiction prose as closely as we now study poems, novels, and plays" (349). A certain highpoint in this trajectory is marked by Ross Winterowd's *The Rhetoric of the "Other" Literature*, published in 1990. Later during the 1990s, compositionist defenses of the essay would shift grounds. Kurt Spellmeyer, Paul Heilker, and others would claim the genre's unique abilities to produce knowledge through a mode relatively less fettered by the kinds of protocols and paradigms now assailed by postmodernity. The essay was a genre tolerant—even celebratory—of the situatedness of the author, whose presence and limitations it foregrounded rather than decried. For some, echoing Lyotard, the essay was the ultimate postmodern genre, able to transform into pedagogical forms like Lester Faigley's microethnographies. For others standing differently in relation to critical theory, the essay was a rallying point against the legacies of Barthes, Foucault, and others whom they feared denigrated authorship.

It is possible, in fact, to regard creative nonfiction's recent popularity as a response to theory. After all, genres like the essay depend on and celebrate the focalizing consciousness of the essay narrator, someone able to take on all manner of topics, however large or small, familiar or foreign, and through the force of his or her learning and personality, comment on them or incorporate them with his or her own experience. At the genre's outset, Montaigne could write about cannibals. Charles Lamb could later write about roast pigs, Virginia Wolf about London air raids, Joan Didion about the Sharon Tate murders, in a genre that Edward Hoaglund claims exists on a line between "what I think and what I am": a person able eloquently to transform event and idea into deft prose.

Consider the different writing traditions converging in the late 1980s. There was a composition tradition born from the late nineteenth-century effacement of rhetoric by exposition in American universities. There was a composition tradition born out of literary studies, a tradition maintained by people interested in the production of creative nonfiction, certainly, but mainly as those genres participated in a Romantic aesthetic. There was, third, a rhetorical tradition not much interested in creative nonfiction at all except as it marked the academy's unfortunate drift from persuasive discourse or, from the perspective of ideological critique, the academy's embrace of bourgeois values. There was, finally, a creative writing tradition birthed by composition, with one godparent twentieth-century literary studies, a tradition interested in the aesthetic but decidedly more from the vantage of production than interpretation or critique. This mix offers a number of Burkean ratios for analysis, but the most interesting ones, especially for a volume proposing *The Spaciousness of Rhetoric*, are the relationships between the rhetorical and the aesthetic and between composition and creative writing.

DEEPER HISTORY LESSONS: EIGHTEENTH-CENTURY PARALLELS

The relationships between the rhetorical and the aesthetic are hardly new. In 1783 Hugh Blair's *Lectures on Rhetoric and Belles Lettres* observed that

> Logical and ethical disquisitions . . . point out to man the improvement of his nature as an intelligent being. . . . Belles Lettres and criticism chiefly consider him as a Being endowed with those powers of taste and imagination, which were intended to embellish his mind, and to supply him with rational and useful entertainment. . . . They bring to light various springs of action which without their aid might have passed unobserved; and which though of a delicate nature, frequently exert a powerful influence on several departments of human life. (10)

Blair's interest in the powers of "taste and the imagination" have a particular resonance after theory that has so thoroughly problematized the politics of tradition and the canon. Blair focuses considerably less on genre than on style, analyzing sentence types, tropes, and figures in over a dozen of the lectures. Given the historical place of style among the canons of rhetoric, the emphasis appears relatively conservative. Yet in his selection of illustrations and texts for analysis, Blair strikes newer ground, devoting four lectures to analyzing Joseph Addison's style in *The Spectator*. More interesting than Blair's method is his applying it to periodical essays rather than more explicitly rhetorical performances. In choosing them, he turned to a comparatively new genre. The first English magazines began only in 1690—among them were the penny weekly *The Athenian Mercury* (1690-97); *The Gentleman's Journal*, a monthly blend of news, prose, and poetry (1692-94); and *The London Spy*, which contained a running narrative of London life (1698-1700). While there were fledgling newspapers, mainly publishing foreign and financial news, during the seventeenth century, English newspapers didn't receive their start in earnest until the more famous publications of Daniel Defoe's *The Review* (1704-13), Richard Steele's *The Tatler* (1709-11), and Addison and Steele's *The Spectator* (1711-12). These last were like newspapers more in terms of their frequency of publication (daily to thrice weekly) than in terms of their content. *The Tatler* and *Spectator* included humorous character sketches ("Ned Softly," "The Political Upholsterer," "Tom Folio"), philosophical essays, reports on the fictional "Spectators Club" member activities, and ruminative personal essays like Addison's "Thoughts in Westminster Abbey." These were entertainments, often but not always with a particular point, but rarely with a specific rhetorical exigency. Other papers and magazines followed, featuring a similar content and adding literary criticism, most famously Samuel Johnson's

The Rambler (1750-52) and *The Idler* (1758-60), and Oliver Goldsmith's *The Bee* (1759).

My point is that when Blair began composing his essays in 1759, new genres were emerging through—and as a result of—new publishing forms. Recall that Blair was a contemporary not only of Johnson but also of Boswell, his biographer. Blair tried to account for some of the genre complexities in his Lectures, dividing between poetry and prose, the latter works divided into "Orations, or Public Discourses of all kinds, . . . Historical Writing, Philosophical Writing, Epistolary Writing, and Fictitious History" (II: 259). There are obvious problems with this division. (Is Emerson's "The American Scholar" oration or philosophy? Richardson's *Pamela* epistolary writing or fictitious history?) Part of the current difficulty in assigning ownership to creative nonfiction comes from the subtle ways categories like Blair's still shape our thinking about prose, especially in those writers like Woolf, Orwell, or Loren Eiseley who philosophize through narratives written in the tone of letter-writers to intimate friends.

Blair counted three "inferior subordinate species" of historical composition: annals, memoirs, and lives, with the memoir being particularly interesting in relation to creative nonfiction (II: 261). In his view:

> Memoirs denote a sort of Composition, in which an Author does not pretend to give full information of all the facts respecting the period of which he writes, but only to relate what he himself had access to know, or what he was concerned in, or what illustrates the conduct of some person, or the circumstances of some transaction, which he chooses for his subject. From a Writer of Memoirs, therefore, is not exacted the same profound research, or enlarged information, as from a writer of History. He is not subject to the same laws of unvarying dignity and gravity. He may talk freely of himself; he may descend into the most familiar anecdotes. What is chiefly required of him is, that he be sprightly and interesting; and especially, that he inform us of things that are useful and curious. . . . This is a species of Writing very bewitching to such as love to write concerning themselves. . . . (II: 286)

With this characterization Blair anticipates many of the features not only of contemporary memoirs but also of creative nonfiction generally: the acknowledgement, even celebration, of the author's limited, interested perspective; the prizing of stories and anecdotes; the injunction to be engaging and interesting. Also current is the careful line that Blair walks between two purposes of belles-lettres: to inform and to entertain. Yet he suggests a third purpose. Recall his earlier assertion that such discourses are "intended to . . . supply . . . rational and useful entertainment. . . . They bring to light various springs of action which without their aid might have passed unobserved" (10). One might regard "bringing to light various springs of action" simply as a sort of inventional work furnishing materials for more centrally rhetorical acts. Or one might regard creative nonfictions as rhetorical performances in their own right, persuading readers to see and interpret the world in specific ways. The same might be said, of course, for poetry, to which Blair devotes several lectures. But nonfiction, grounded in the problematic "real" and "true," stands closer to rhetoric's traditional domain. Creative nonfiction might be located within the broad category of epideictic discourse, a category that Cynthia Sheard explains needs to be understood in terms broader than Aristotle's rhetoric of praise and blame. Sheard reviews arguments for the epideictic's poetic tendencies, its artistic effects, its literary qualities, and its linkage with ethos and pathos (774). If composition studies wished to claim creative nonfiction, its best two-pronged gambit might be to position composition in rhetoric and creative nonfiction in the epideictic as that nonfiction prose most depending on ethos and pathos. However, even though such moves are plausible, they strike me as rather more motivated by taxonomic tidiness than by conceptual rigor.

OUT OF HISTORY

Such moves would invoke a kind of rhetoric of priority in which current "ownership" is determined by previous historical claims. If a major historical figure like Hugh Blair can be shown at least to affiliate with rhetoric those genres we now understand as creative nonfiction, then rhetoric and composition studies has the deed to them, the pedagogical rights, the research rights, the mineral and water rights, and so on. Through this rationale, Scholes, Berlin, Donald Stewart, and others have asserted the priority of rhetoric, not literature, in English departments that developed out of nineteenth-century oratory. However, the rhetoric of priority works better as heuristic (or self-righteous justification) than as sufficient argument. Corder noted that to insist on rhetoric as defined by the ancients (or even the near ancients) "is tantamount to accepting a Ptolemaic cosmology, repudiating Newton's work on gravity, dodging quantum physics, mowing one's lawn with a scythe, and using ground-up-black-eyed peas for coffee" (332).

I hope that this brief history shows how complicated the position of creative nonfiction has been and continues to be in the academy. At the outset I asked, "Who owns creative nonfiction?" At some level it really doesn't matter, as long as whoever claims it isn't selfish. I must admit being peeved at some who dismiss rhetoric and composition without having much read in the history of these fields, their theories, pedagogies, and institutional histories. They can't smugly imagine that, to paraphrase Ann Richards, creative nonfiction stood on third base in the 1990s and creative writing programs hit a triple to put it there.

But at another level, it does matter who owns creative nonfiction—or at least who would claim to own it. In this respect I want to critique rhetoric and composition studies and its telling failure to grapple with issues represented by my opening question.

The competition over creative nonfiction underscores a continued disciplinary fragmentation of writing studies. In 1987 Jim Corder observed that a longstanding two-part division between literary studies and rhetorical studies had developed into a three-part division, with composition "now acquiring methodology and research of its own, on point of becoming a third segment, no longer fully connected to the work of rhetoricians" (334). Writing in English studies has since further splintered into at least fourth and fifth segments: technical/professional writing and creative writing. Obviously, both these traditions have existed programmatically for some time, although as I have indicated above in my brief review of creative writing's history, not nearly as long as we might think.

And while we're considering these programmatic histories, we might also consider some largely unacknowledged assumptions each makes about writing. Technical writing claims the vocational applications of exposition in which texts, however instrumental, are means through which various kinds of work happen. Creative writing claims the identity of writer, cultivating an ethos of the student not as "English major who writes" or "biologist who writes" but as writer who writes, the writing being its own ends. In this respect, technical writing in the academy is surprisingly like creative writing—and like journalism. (Richard Lloyd-Jones provides a rich vignette of Iowa Writers Workshop poets in the 1950s teaching technical writing to engineers.) All these fields offer an ultimate prospect of writing for income, though I scarcely claim the likelihood of this prospect, especially for poets. In contrast, composition claims . . . what? Rarely is it writing as an end in itself. Instead composition often focuses on matters of academic form (thesis-and-support structures, discourse conventions, etc.). Composition focuses, too, on individual development, cultural critique, or argument. I overstate to make a point, but composition regards writing as a means of doing in the world more than as a way of being in the world, as a tool for achieving everything from knowledge to political action. To redirect James Britton's formulation from the reader, composition foregrounds the writer as participant rather than as spectator. I want to be careful here, because I'm getting close to an argument for Arnoldian disinterestedness as a justification for writing, and what I want to argue instead is an expansion of composition's sphere.

Composition has both benefited and suffered under its status as servant. I hardly think providing service is a bad thing. I'm both pragmatic and altruistic about the values of writing for students' economic or civic participation. I certainly acknowledge composition's economic value to English departments and universities—if not always to writing programs themselves. But as composition remains reduced to fresh-year comp or writing across the curriculum and valued almost solely for its instrumentality, it oscillates between a nineteenth-century academic rhetoric of perspicuity and a vestigial notion of rhetoric as civic discourse, modified occasionally by writing as personal development. As a result, composition stores writing mainly in two sites, one academic and one political, although a limited politics in which discourse always exists in relation to clear exigencies. Technical, business, or professional writing handle the vocational sphere, and creative writing the leisure sphere—or at least some of it—and journalism handles news and reportage, including texts not limited by timeliness: "features."

But. There is a realm of popular discourse not clearly claimed by these entities, a domain expanding since printing and modernism begat several now-three-centuries-old genres, in periodicals and books neither expressly didactic nor explicitly argumentative, texts that are not necessary in any direct sense. Who needs to know Annie Dillard's Tinker Creek or Michael Martone's Iowa farmsteads? These are texts rather read in leisure, with that term understood in its broadest and most profound dimensions, the very act of reading providing pleasure of experience and ideas made interesting through language. Given that the cultural role of these texts parallels the role played historically by fiction, poetry, and drama, it is perfectly reasonable that creative writing programs would claim some of them.

The crucial question is whether composition now much claims genres other than those that live mainly in the academy. I think mostly not. Composition studies is more concerned with writing rather than with writings. It supports identities of "students as writers" or, say, "biologists as writers," subject positions that subordinate "writer" to some prior and primary identity. Composition studies does not generally support the complementary subject positions of "writer as student" or "writer as biologist," in which the subject position of writer is foregrounded. One quality occluded in composition's very important political and social turns is that of writing as craft, as the making of textual artifacts whose maker is important *as* maker. Articulating a relationship between creative nonfiction and composition studies would help to inscribe that subject position, not as an exclusive one but certainly as a vital one.

Teaching creative nonfiction is certainly not the only way to do this. Composition programs should be wary of narrow identities as sites for generic composition courses. In addition to first-year comp and instead of generic advanced comp, we might imagine specific courses in "Civic Argument," "Autobiography and Memoir," "Reporting," "Research and Direct Observation," "Fiction," "Documentary Writing," "Writing for Cyberspace," "Academic Issues and Popular Audiences," "Travel Writing," and so on. Since some of these courses already exist, if not in composition programs in journalism or creative writing, this may seem to advocate a composition land grab. Rather I'm interested in articulation among programs, an effort perhaps best occasioned by the challenge of jointly developing a coherent set of courses that represent the range of writing and genres. There are interesting curricular experiments toward these ends. For example, the Department of Writing and Rhetoric at the University of Central Arkansas combines rhetoric, composition studies, technical writing, and creative writing in a freestanding department located, tellingly, in the College of Fine Arts.

Composition is central in these curricular endeavors because of its status in the academy, its importance reflected in universal requirements, its bipolar unimportance reflected in the funding and staffing of those requirements. This centrality means that composition courses represent writing for most students. In the vital aspects of academic discourse and argument, they do so well. But writing is a broader category than composition, and the question is how broadly we should represent writing to students. We should at least ask what it portends when genres become so academically segregated that student writers—or their professors—do not perceive and wonder about the complex relations among them.

It is just as debilitating for compositionists to snub creative writing for holding undertheorized views of writing as it is for creative writers to snub composition for merely transmitting rudiments. The genres of creative nonfiction, at least for now, inhabit a kind of middle ground between composition and creative writing programs. Creative writing currently—and appropriately, I finally think—assumes the most visible caretaking role for these textual commons, which had fallen into relative neglect. A more expansive view of rhetoric and composition, appropriating Blair's work in ways such that the academic subject would have emerged as writing (encompassing both rhetoric and belles-lettres) rather than as composition, may have settled the title of these lands differently. The challenge is for creative writing programs to understand why rhetoric and composition has a continued important stake in these fields, one important not only for historical and conceptual reasons but also for the ways large numbers of students understand the terrain of writing and their own possibilities as writers. The even bigger challenge is for composition programs to understand this stake too.

WORKS CITED

38th Annual CCCC Convention Program. Urbana, IL: NCTE, 1987.

Anderson, Chris. "Literary Nonfiction and Composition." *Literary Nonfiction: Theory, Criticism, Pedagogy.* Ed. Chris Anderson. Carbondale: Southern Illinois UP, 1989. ix-xxvi.

Babcock, Robert Witbeck, Robert Dewey Horn, and Thomas Hopkins English. *Creative Writing for College Students.* New York: American Book, 1938.

Berlin, James. *Rhetorics, Poetics, and Cultures.* Urbana, IL: NCTE, 1996.

Blair, Hugh. *Lectures on Rhetoric and Belles Lettres.* 2 vols. Ed. Harold F. Harding. Carbondale: Southern Illinois UP, 1965.

Brereton, John C., ed. *The Origins of Composition Studies in the American College, 1875-1925: A Documentary History.* Pittsburgh: U of Pittsburgh P, 1995.

Britton, James. "Spectator Role and the Beginnings of Writing." *Prospect and Retrospect: Selected Essays of James Britton.* Ed. Gordon M. Pradl. Montclair, NJ: Boynton/Cook, 1982. 46-70.

Brooks, Stratton D., and Marietta Hubbard. *Composition-Rhetoric.* New York: American Book, 1905.

Butrym, Alexander J., ed. *Essays on the Essay.* Athens: U of Georgia P, 1990.

Cheney, Theodore A. Rees. *Writing Creative Nonfiction.* Berkeley, CA: Ten Speed P, 1991.

Connors, Robert J. *Composition Rhetoric: Backgrounds, Theory, and Pedagogy.* Pittsburgh: U of Pittsburgh P, 1997.

Corder, Jim W. "Studying Rhetoric and Literature." *Teaching Composition: Twelve Bibliographical Essays.* Ed. Gary Tate. Fort Worth: Texas Christian UP, 1987. 331-52.

Department of Writing and Rhetoric, University of Central Arkansas. http://www.uca.edu/divisions/academic/writing/. (11 Jan. 2001).

Dillard, Annie. "Introduction." *The Best American Essays 1988.* Ed. Annie Dillard. New York: Ticknor and Fields, 1988. xiii-xxii.

Faigley, Lester. *Fragments of Rationality.* Pittsburgh: U of Pittsburgh P, 1992.

Fenza, David. "A Brief History of AWP." http://awpwriter.org/history.htm. (20 Sept. 2000).

—, ed. *The Official AWP Guide to Writing Programs, 9th Edition.* Paradise, CA: Dustbooks, 2000.

Gerard, Philip. *Creative Nonfiction: Researching and Crafting Stories of Real Life.* Cincinnati: Story P, 1999.

Heilker, Paul. *The Essay.* Urbana, IL: NCTE, 1996.

Lloyd-Jones, Richard. "Poesis: Making Papers." *Writing on the Edge* 8 (Spring/Summer 1997): 40-46.

Lounsberry, Barbara. "Nonfiction Writers Mean to Have It All." *North American Review* 285 (Sept./Oct. 2000): 43.

Martone, Michael. *The Flatness and Other Landscapes.* Athens: U of Georgia P, 2000.

Myers, D. G. *The Elephants Teach: Creative Writing Since 1880.* Englewood Cliffs, NJ: Prentice, 1996.

Rose, Mary. (Associated Writing Programs.) Telephone conversation. 2 November 2000.

Scholes, Robert. *Textual Power: Literary Theory and the Teaching of English.* New Haven, CT: Yale UP, 1985.

Sears, Donald A., ed. *Directory of Creative Writing Programs in the United States and Canada, Second Edition.* Fullerton, CA: The College English Association, 1970.

Sheard, Cynthia Meicznikowski. "The Public Value of Epideictic Rhetoric." *College English* 58 (Nov. 1996): 765-94.

Spellmeyer, Kurt. "A Common Ground: The Essay in the Academy." *Common Ground.* Englewood Cliffs, NJ: Prentice, 1993.

Stewart, Donald. "Two Model Teachers and the Harvardization of English Departments." *The Rhetorical Tradition and Modern Writing.* Ed. James J. Murphy. New York: MLA, 1982. 118-29.

Winterowd, W. Ross. *The Rhetoric of the "Other" Literature.* Carbondale: Southern Illinois UP, 1996.

Author Index

Subject Index

A

Argument as charm, 66–67
Argument as bold assertion, 68–70
  ethos in persuasion, 69
  rhetoric is an architectonic discipline, 68–69
  truth is jacklegged, 69–70
Argument as incompleteness, 67–68
Argument as indirection, 60–65
  bouncing, 62
  building clotheslines, 61–62
  circling, 60–61
  exploding taxonomical boxes, 63–64
  refusing to theorize, 64–65
  reveling in self-contradiction, 62–63
Argument as mystery, 65–66
Argument requires love, 70–72
Aristotle on civility, 136–38, 139, 142 (*see* Rhetoric and incivility)
  civic virtue, 150
  morality, 137–39
  virtue, 137–39, 142
Aristotle's definition of rhetoric, 236

C

Civic, 139, 148
Civil, 139–40, 182
Civilizing process, 139–42
  early societies, 140–42
Classical rhetorical terms, 108
  narratio, 109–10
  oratio, 108–09
Comity, 131, 135, 144, 154
Composition instruction, 6–16, 85
  disciplinarity, 9–10, 14–16
  evaluation, 13

  invention and audience, 6–8, 11
  pedagogies, 9, 13–14
  studies of styles, 8–9, 12
  theories of discourse, 9, 11–12
Conflict, 172–77, 195
  political, 175–76
Conflict resolution, 173–80
  in literary theory, 173–80, 182
  international peacemaking, 173–74
  rhetoric of problem-solving, 175–77
Compressed time and space, 133, 144
Constructing and reconstructing, 148
Corder vs. Kinneavy, 119, 126–27
Corder's personal-geography course, 235, 236, 247
  home, 243–45
Corder's personal narrative, 38, 92–94, 98, 108–15, 117, 127
  language of West Texas, 104
Corder's research, 105, 106, 114, 115, 122, 123
  search for nostalgia, 106, 107, 122, 123
Corder's rhetoric, 150–54
Corder's rhetorics of war, 192–96
  diaspora, 193, 194
  discomposed self, 192–96
  invention, 193–95
  testimony, 195, 196
  Yonder, 193, 194
Corder's scholarly work invoking personal, 108–15
Corder's style and identity, 89–99
  essay titles, 96, 100, 101
  his identity with his community, 90–91, 96
  love in academic essays, 97
  sentence style, 99–100
Corder's search for identity, 122–24

271

NOTES ON CONTRIBUTORS

James S. Baumlin is Professor of English at Southwest Missouri State University, where he teaches English Renaissance literature and the history of rhetoric. His publications include *John Donne and the Rhetorics of Renaissance Discourse; Ethos: New Essays in Rhetorical and Critical Theory*, coedited with Tita French Baumlin; and *Rhetoric and Kairos: Essays in History, Theory, and Praxis*, coedited with Phillip Sipiora.

Wendy Bishop, Kellog W. Hunt Professor of English at Florida State University, teaches composition, rhetoric, poetry, and essay writing. A former writing center director and writing program administrator, she studies writing classrooms, writes assignments with her students, and shares her work in *Teaching Lives; Elements of Alternate Style; Thirteen Ways of Looking for a Poem; Ethnographic Writing Research; The Subject Is Research; Reading into Writing, A Guide to Composing;* and *The Writing Process Reader.* She lives in Tallahassee and Alligator Point, Florida, with her husband Dean and children Morgan and Tait.

Peter Elbow is Emeritus Professor of English at the UMass Amherst, where he directed the writing program. He taught at MIT, Franconia College, Evergreen State College, and SUNY Stony Brook, where he also directed the Writing Program. He is author of *Writing without Teachers, Oppositions in Chaucer, Writing with Power, Embracing Contraries, What Is English?* His most recent book, *Everyone Can Write: Essays Toward a Hopeful Theory of Writing and Teaching Writing*, was given the James Britton Award by the Conference on English Education. His textbook with Pat Belanoff, *A Community of Writers*, is in its third edition. NCTE recently gave him the James Squire Award "for his transforming influence and lasting intellectual contribution to the English Profession."

Theresa Enos is Professor of English and Director of the Rhetoric, Composition, and the Teaching of English Graduate Program at the University of Arizona. Founder and editor of *Rhetoric Review*, she teaches both graduate and undergraduate courses in writing and rhetoric. Her research interests include the history and theory of rhetoric and the intellectual work and politics of rhetoric and composition studies. She has edited and coedited nine books, including the *Encyclopedia of Rhetoric and Composition* and *The Writing Program Administrator's Resource: A Guide to Reflective Institutional Practice*, and she has published numerous chapters and articles on rhetorical theory and issues in writing. She is the author of *Gender Roles and Faculty Lives in Rhetoric and Composition* (1996) and past president of the national Council of Writing Program Administrators (1997-99).

Elizabeth Ervin is Associate Professor of English at the University of North Carolina at Wilmington where she teaches courses in writing and literacy, English education, and women's studies. Her recent research explored the dynamics of public discourse, including the question of why people so often misunderstand and vilify each other when discussing issues of mutual interest. Writing the chapter featured in this collection reminded her of how fortunate she is to be in a profession where she can ponder the fundamental questions of her heart in the company of patient, supportive colleagues.

Doug Hesse is Wiepking Professor at Miami University (Ohio) and Professor of English at Illinois State University. He served as president of the Council of Writing Program Administrators from 1999-2000, edited *Writing Program Administration* from 1994 to 1998, and is a member of the CCCC Executive Committee. Educated in small-town Iowa and at the University of Iowa, he has received distinguished teaching and research awards from Illinois State. Among his publications are articles in *College Composition and Communication, JAC, Rhetoric Review, Writing on the Edge, Journal of Teaching Writing, The New England Journal of Medicine*, and chapters in *Questioning Authority; The Writing Program Administrator's Resource; Passions, Pedagogies, and 21st Century Technologies; Writing Theory and Critical Theory; Essays on the Essay; Narrative and Argument; Literary Nonfiction;* and others.

Pat C. Hoy II, Director of the Expository Writing Program and Professor of English at New York University, has also held appointments as Professor of English, US Military Academy, and as senior preceptor, Expository Writing Program, Harvard University. He received his BA from the US Military Academy and his PhD from the University of Pennsylvania. Professor Hoy is the author of *Reading and Writing Essays: The Imaginative Tasks* and *Instinct for Survival: Essays by Pat C. Hoy II*. He is the coeditor of *Encounters: Essays for Exploration and Inquiry* and coauthor of *The Scribner Handbook for Writers* (with Robert Di Yanni). His essays on pedagogy appear in *Literary Nonfiction: Theory, Criticism, Pedagogy; How Writers Teach Writing;* and *What Do I Know?: Reading, Writing, and Teaching the Essay*. Other essays and reviews have appeared in *Agni, Rhetoric Review, Sewanee Review, South Atlantic Review, Twentieth Century Literature, Virginia Quarterly Review*, and *Writing on the Edge*. Six of his familiar essays have been selected as "Notable" in *Best American Essays*.

Janice M. Lauer is the Reece McGee Distinguished Professor of English at Purdue University, where she founded and directed the graduate program in Rhetoric and Composition. For thirteen years she directed a national summer Rhetoric Seminar. She has coauthored *Four Worlds of Writing: Inquiry and Action in Context*; *Composition Research: Empirical Designs*; and articles on invention, rhetorical history, disciplinarity, and pedagogy. She received the CCCC Exemplar Award and chaired the NCTE College Section and the MLA Division of the History and Theory of Rhetoric and served on the executive committees of CCCC, NCTE, and the board of directors of The Rhetoric Society of America.

Richard Lloyd-Jones retired from the University of Iowa (1952-95). Former chair of the Conference on College Composition and Communication, first recipient of their Exemplar award, former President of NCTE, he received the second Francis Andrew March Award for professional service of the Association of Departments of English. Lloyd-Jones is author of various articles on professional issues including coediting the reports of the English Coalition Conference and is codesigner of Primary Trait Scoring.

Keith D. Miller is the author of *Voice of Deliverance: The Language of Martin Luther King, Jr. and Its Sources*, which was issued in a second edition by University of Georgia Press. He has also published numerous scholarly essays on King, on the songs of the civil rights movement, and on Frederick Douglass. He recently coedited *New Bones: Contemporary Black Writers in America*, a large multigenre anthology of African-American literature. For two years he served as Writing Program Administrator at Arizona State University.

John Warnock teaches at the University of Arizona where he directed the graduate program in Rhetoric, Composition, and the Teaching of English from 1992 to 1997. He taught at the University of Wyoming from 1970 to 1991. To help him make his point, he has included a good deal more biographical information in the essay for which this is the biographical note.

Tilly Warnock is Associate Professor of English and Director of the Composition Program at the University of Arizona. She has published articles on rhetoric and composition and a composition textbook, *Writing Is Critical Action*, as well as coedited with Joe Trimmer *Understanding Others: Cultural and Cross-Cultural Studies and the Teaching of Literature*.

W. Ross Winterowd is the founder and was the long-time director of the doctoral program in Rhetoric, Linguistics, and Literature at the University of Southern California. That program is now, lamentably, defunct, but more than fifty graduates throughout academia carry on the work begun at USC. Winterowd's most recent work is *The English Department: A Personal and Institutional History*.

George E. Yoos, Emeritus Professor, St. Cloud State University, taught philosophy for thirty-two years. Prior he taught science and was principal of a private preparatory school in Chicago. His early work in philosophy was in informal logic, aesthetics, figurative language, and pragmatics, which led him into rhetoric and his early involvement with the Rhetoric Society of America. He served as editor of *Rhetoric Society Quarterly* from 1972 to 1990. His special interests in rhetoric are critical thinking, pictorial rhetoric, religious and political rhetoric, ethos and ethical appeal, and especially about what he distinguishes as a rhetoric of appeal and a rhetoric of response.

Richard E. Young is Baker Professor of English, Emeritus, at Carnegie Mellon University. Among his many contributions to rhetorical studies are "Toward a Modern Theory of Rhetoric: A Tagmemic Contribution" (with Alton Becker, *Harvard Educational Review* 36, 1965), *Rhetoric: Discovery and Change* (with Alton Becker and Kenneth Pike), "Arts, Crafts, Gifts, and Knacks: Some Disharmonies in the New Rhetoric" (*Reinventing the Rhetorical Tradition*), "Why Write?: A Reconsideration" (with Patricia Sullivan, *Essays on Classical Rhetoric and Modern Discourse*), "Recent Developments in Rhetorical Invention" (*Teaching Composition: 12 Bibliographic Essays*), and *Reading Empirical Research Studies: The Rhetoric of Research* (with John R. Hayes et al.).